GW01311084

Contents

Dispatch No 1 – Monday, January 7

I didn't ask because I give a fuck

I'm due back at work on the mining site in two days, but haven't been sent any flight details. Maybe I was fired over the break and nobody told me. Five of the guys have texted me over the last few days wondering about their flights as well, however, so unless we're all done or JRT has gone bust, I should still have a job.

While waiting for a train to my girlfriend's place, I call my boss onsite, Jonno. As the phone rings, I realise I should've made this call a week ago, but that's how someone who's after a career would behave, not a trying-to-be comedian desperate for cash.

'What's cracking? How are you, chief?' he asks.

'I'm good. And yourself?'

'Kicking goals from everywhere. How was your Xmas?'

'Really good. Did some of my biggest ever comedy shows,' I reply.

'So you finally had tens of people at a show, instead of just single digits? Onya tiger.'

'More like a couple of thousand. It was amazing, and I got to meet...'

'Let me stop you there and make something very clear. I didn't ask because I give a fuck.'

It takes me a moment to remember that Jonno might not be being mean and, even if he is, it's pointless arguing. Nobody has spoken to me like this since I last saw my fellow mining site workers, at the airport after we arrived home for the beginning of the break and Damo told me with a smile, 'I hope you get herpes for Xmas. And not the face ones. It's what you deserve.'

'How was your break?' I ask Jonno.

'Fucking nearly over now.'

'Oh, okay. So I don't have any flight details yet.'

'All under control chief. You'll have them either today or tomorrow at the latest,' he replies.

Ten minutes later I'm on the train, and Jonno phones back.

'Right Xavier, I need you to organise everyone's flights.'

So I phone Debitel, the company managing the site where we're working.

They're responsible for booking all of the contractors' flights, but only after we place a request, which is supposed to be made at least a month in advance. Not two days, and I'm not even sure if they're back onsite yet.

'Debitel. Samantha speaking.'

'It's Xavier here from JRT Projects. When did you guys start back?'

'January second,' she replies.

'Wow. That's not much of a break.'

'No, it's not.'

From her briskness, I can tell how excited she is to be back living the FIFO life.

'FIFO' stands for 'fly-in-fly-out', and describes how most of the workers get from the Australian cities where they live to the remote locations where the mining happens. Well, that's the official meaning, but I was originally told it means 'fit in or fuck off'. There are also 'DIDOs' who drive in and drive out from the mining site, and I've always wished there was some way to add an extra 'L' into that acronym. Then I imagine a dildo trying to drive a car. How would that even work? It'd need legs to reach the pedals, and adorable little dildo arms to steer.

'So I'm not sure if they've been booked or not, but none of our guys have their flight details,' I tell Samantha.

'You should have them, and you're due back here on Wednesday, right? If you haven't put in a request, you're stuffed. The flights are booked solid.'

'Well, we might have a very big problem,' I reply through a smile, as I might get extra time off.

'Give me a second.'

I hear the handset hit the desk. A minute later it rustles, then she's back.

'The flights were booked last year. You even signed the request. You don't remember?'

'Nup.'

My plan was to think as little as possible about work over the break, and it seems I've succeeded.

'We sent through a confirmation email to Jerome. Does he even still work for JRT?'

'As far as I know.'

She pauses. 'Why?'

'Honestly, I have no idea.'

'I thought you would've canned him ages ago.'

‘So can you please forward me that email? That you sent to him?’ I ask.
‘Why can’t you just get it from Jerome?’ She laughs. ‘I’ve just sent it.’
Jerome is the other onsite JRT Projects admin person, Jonno is the boss, Dale is the supervisor and there are about thirty assorted tradesmen. Across the site where we’re working, the various companies employ around twenty admin people, and Jerome and I are the sole males. I still remember the looks of disdain I got on my first day from every man, and I thought it was just because they didn’t like me, but those initial looks were more to do with the fact that I’d replaced Jessica. In the words of Jonno, ‘She was so hot that even if she was my first cousin, that wouldn’t stop me.’

Over the break, during one of the rare times I’d remembered that I had a day job, I’d decided to give Jerome a chance. It’s a fact that most inanimate objects are both more useful and have a higher IQ, but he’s a human being who I’m sure has good intentions. Somehow, someone at some stage had seen enough to give him a job, and throughout his working life a series of employers have done the same, which means he has to be capable of something. The problem is that we’ve all been expecting too much, and I need to tailor his tasks to his abilities. Now I’m still going to give him a chance, but we haven’t even started back and he’s already causing havoc.

I forward the flights to Jonno, and follow up with a phone call, as I know he won’t check his email unless prompted.

‘Instead of me emailing all the guys one at a time,’ he says, ‘why don’t you do that?’

‘Sure. By the way, apparently they were sent to Jerome last year.’

‘I can’t believe that thing still has a job,’ Jonno replies.

‘Wait, aren’t you in charge of that sort of thing?’

‘Jerome was hired by Peter. So only Peter can fire him. See you Wednesday, sweetheart.’

Peter and Scott are JRT’s two directors, and last year I learnt that Peter likes to ambush employees by yelling at them over minor and made-up problems, but prefers to avoid genuinely awkward encounters. So he often organises others to do the firing for him or, as with Jerome, ignores the situation entirely. Being in charge of a company with tens of millions in annual turnover, however, it’s possible he just hasn’t got around to Jerome yet, as he’s got more important things to do than monitoring the output of someone without any.

An hour after arriving at my girlfriend's house because, you know, sex, I use her computer to email the flight details to the JRT employees, then send follow-up text messages. Most reply immediately, and I'm reminded again how quickly these men respond when it's information they're chasing, but how elusive they can be when you want something from them.

We work three weeks on the mining site, then get one week off, and that week off is staggered to leave a certain number of employees onsite at all times. Four of the guys whose week off is that first week back call me to complain that they can't last another week without wages. They all make around $3,000 per week after tax, but those five on an enforced extended Xmas break won't be making anything, as these men aren't on a salary but are paid per hour worked. Instead of attempting to placate them, I tell them to take it up with Jonno, which they won't, as we all know they'd be told to 'shut the fuck up or find another job'.

Other than the calls, everyone texted single word replies, apart from:

Donk, the safety guy, who sent me a photo of two naked women pleasuring a horse.

Pando, who sent the same horse photo.

Damo, the machine operator, who texted, 'Thanks. Herpes?'

Roger then went with, 'Roger that.'

Jerome said, 'I already knew that, but thanks anyway!'

Dale was one of those who had an extra week, and texted, 'Seven days and counting, ball-bag.'

Then Fatty sent a picture of a hot-looking woman with a dick that read, 'Everything was perfect. Until you cocked it all up.'

Dispatch No 2 – Tuesday, January 8

Bob Hawke walked out of my show

Tomorrow I fly out to the mining site, and tonight I'm having dinner with my parents because I still live at home and what of it? Sure I'm thirty-four, but what's the point of paying rent somewhere, or a mortgage, when I'm only going to be there seven days out of every four weeks?

Right now though, I'm laying in my girlfriend's bed drinking a beer, and winning at life. Which has a very big chance of changing, as of tomorrow.

I haven't done anything that special with this last day, but it all feels special because it's going to be a while before I get to do any of it again. I've never been to jail, but I imagine right now feels a little like the night before you're put away.

Similar to actually being in prison, from tomorrow I'm going to be trapped in the middle of nowhere for an extended period surrounded by a high number of men all tougher than me. The downside of my situation is that I'll have less spare time than I would in jail, as I'll actually have a job to do. Leaving far less time for reading and writing, which in my jail fantasies is all I'd be doing, apart from exercising and having intellectual conversations with the guards and fellow inmates, and not being raped.

I've never looked into the reality of jail, so I suspect I'm way off.

Maybe war is a better comparison, but I've never done that either and, at the mining site, nobody will be trying to kill me. Well I hope not, but the mining site is in the outback, I've spent a large part of my past as a backpacker, and I have just watched *Wolf Creek* 1 and 2.

Perhaps camping is the best comparison? Where I'll be staying is called a campsite, but there's no hiking, natural wonders or weird European families whose clothing is either way too short or long. Whenever I've been camping they're always there, with their big families, in which it's impossible to tell who's related to whom and how, and it's always the men wearing too little, and the women too much. They also often have a better knowledge of English than I do.

'Excuse me, kind sir? Do you know how to get to the isthmus? Is it beside the estuary, or closer to the mesa?'

So I suppose you could say that I'm going off to a short war with no enemy, or camping without Europeans, or jail without the actual imprisonment, and with far better pay.

What it's most like is a huge building site. Where you're stuck for three weeks at a time, twelve hours a day. Then at the end of every day, you don't get to go home but have to stay there, with only a TV and the same people you've been working with all day for company, and over 95 per cent of them are men. That's mostly what it's like, because that's exactly what it is.

Maybe it'd help if I mentioned where I was going?

Sorry, not going to happen. This book is not about getting anyone I worked with in trouble, and I'd like to be able to mention any illegal things that do happen. During my first three months I witnessed plenty, and these included: working drunk, calling an extended stay in hospital 'light office duties', forging all sorts of documents, drug deals and prostitution, and driving onsite without the proper qualifications, and I'm pretty sure it's only that last one that gets you fired on the spot.

Really, this is just an honest account of one person's short time as a FIFO worker. Holding nothing back, and with no rose-coloured anything, apart from the things that are supposed to be rose-coloured.

I can be honest because if and when anyone reads this, I'll hopefully be long finished working out there. So far I've done three months and my plan is to last around another six, but it could easily be much shorter than that.

So here's the thing. I don't want this job, but I need it. A lot more than I wish I did.

Last year I started as a FIFO, after spending a month performing as a comedian at the 2012 Edinburgh Fringe Festival and amassing $20,000 in debt. Basically, I'd gambled that plenty of people would buy tickets to see this unknown Australian comedian, and I lost. Big time.

After three months of FIFO work, and one day before restarting FIFO work this year, I'm still $20,000 in debt.

Whoops.

Whenever I'm out at the mining site, like most, I spend very little. Many of the guys then use those big wages to have a really good time during any time off. Some fly to Asia for a haze of hookers and cocaine, others spend big on booze and expensive toys and, while I'm not that outlandish, I certainly haven't been frugal.

For example, if I had to choose between a delicious but expensive three-course meal and a can of soup, well, not too long ago I didn't have that choice, so of course I'm going to bypass the soup, as well as paying for any friends that join me.

Also, I've committed to performing this year at Perth Fringe World, the Adelaide Fringe Festival, and the Melbourne International Comedy Festival. Meaning all the flights, accommodation, insurance, registration fees, advertising and marketing, and all the other stuff that has a price, has already been paid for. It's all cash that needs to be laid out months in advance, and hopefully made back through ticket sales, and that money doesn't come in until weeks and sometimes months after the festival is finished.

It's exactly like gambling. Except you need to pay for all your bets upfront, they take months to pay off, and regardless of how hard you work, there's no guarantee that they'll pay anything. In other words, it's gambling for idiots. So that's why I'm here, grinding through the FIFO life, and I'm now unsure how long I'm going to need to stay. In Edinburgh, I went all-in with money I didn't have and lost the lot, and right now, I'm doing the same thing again. Depending on how it turns out, it's either an intelligent choice to back myself, or I'm a moron repeating an identical mistake.

So here's the big problem. My employer, JRT Projects, needs to give me an extra week off here and there in order to perform at these festivals, but nobody has approved my leave.

Leading up to the Xmas holiday I did send emails, make calls and even directly asked my boss, Jonno, but that counts for nothing, because nothing was done. So either they approve my leave and I keep my job – or they don't and this will be a very short book, and I'll still be $20,000 in debt.

Flying home last year from the mining site for Xmas, I was all about choosing my passion over this job if I had to, but I didn't expect to still be down $20,000. Now I certainly haven't changed my mind, but I really hope it doesn't come to that and I can do both. If not, it's likely that there are years more of work that isn't writing or comedy to come, which won't pay anywhere near as well as FIFO work. Which is exactly the opposite of what I thought I was working towards, and makes me feel like I keep making the same mistakes. Or that this whole dream of mine is a mistake.

So now that my holidays are very nearly done, I'm shitting myself, both metaphorically and literally. Seriously, my stomach has been doing actual backflips, with awful results.

Sorry.

That's disgusting.

Over the holidays, on the rare occasions that I spoke to anyone about the FIFO work I was doing, they asked me what it was like, being a coal miner. It's a reasonable question – I am a worker on a coal mining site, but it left me crimson-faced as I admitted that's not what I did, then awkwardly explained what my job actually involved, and that in three months of it I hadn't even met one coal miner.

So this year, if I keep my job, I'm going to search out an actual, real coal miner. If that's even still a thing.

Onsite jobs in mining are broadly split in two – the construction phase and the mining phase. The construction phase hires far more workers, who are employed to build the infrastructure that enables the mining to happen and which includes roads, accommodation and other assorted facilities.

Australia-wide in 2013, which is the year I've just started, there are around 275,000 workers directly employed by the resources sector (the fancy name for mining), and as the intense construction stage shifts to mining site operation, it's estimated that around 75,000 jobs will be lost. Including mine, as I'm in admin for JRT Projects, which is on the construction side.

All the infrastructure is built to last, but not forever. It's going to take a couple of decades or so to get all the coal, and the contract the mining company has with the government states that they must leave the site the same pristine empty as they found it. Provided society hasn't collapsed and human life been extinguished in the meantime.

Australia is the world's second largest exporter of coal and its fifth largest producer, and scientific studies have repeatedly found that if we actually use all the planet's fossil fuels, the result will be an environmental catastrophe. It's estimated by people a lot smarter than me that we have around five times as much oil, coal and gas on this planet as it's safe to burn.

How's this for a comparison? You've got a full wine cellar, and it's all delicious stuff. Then one night, you decide to drink it all. You'll die, but will have likely had an awesome time along the way, apart from the bit at the end where you're vomiting and can't talk or see. That's pretty much exactly what we're doing to the planet, according to climate scientists. Now if you disagree, well, you're wrong, but don't stress out. This book isn't about that. It's meant to be a funny true story, not a tragic one.

Also, I'm not oblivious to the fact that I'm in the wine cellar, shoving bottles into my backpack before rushing out the door to pawn them for whatever I can get. In this case, that's time onstage in a room where I hope I'm joined by paying audience members, but it could turn out that I'm paying upwards of $100 an hour for weeks at a time to rent a room, empty chairs, regret and despair. Really, isn't there at least one way to spend $100 an hour for a room far more wisely? Yes, that's a joke about prostitution. A cheap joke about something that's way out of my price range.

Back to the wine, which in this case is coal, and here's the decision I've made, which I realise is not for everyone, especially those with stronger morals than me. I'm taking the money because, in my opinion, it's hard enough to be an artist these days. Here nobody is telling me what I can and can't say, and without the dirty coal money, I couldn't do the thing I love. Well at least not for years, because I could get a job in hospitality and work far harder for much less money than I'm currently on, but nah.

Oh yeah, and I have a girlfriend now. So, about that. I had been wondering how relationships on mining sites worked, or didn't, and now I get to find out. Last year I did ask some of my fellow FIFOs, but they were as eloquent and expansive about their personal lives and feelings as they are on prohibition and religion. As in, they just didn't talk about it.

Her name's Verity, and we met at the Woodford Folk Festival on New Year's Day, meaning our anniversary will be super easy to remember. She's a lawyer, slightly younger than me and to say any more would be bragging. So let me brag.

We talk, we laugh, the physical stuff is top notch, and we've both previously made enough mistakes to know that when something works and seems easy, it's actually both special and rare.

For the last week of my break, I've pretty much moved in with her, and although we've spent nearly all of every day together, we're still getting on sensationally. I know – spew.

Right now I might be a man in my mid-thirties sounding like a teenager, but that's how I think you should be sounding at this stage. You're either tentative because it's not quite right or you've been hurt before, or you let yourself fall because falling feels great, and although you've been hurt before, this is all worth the risk and is too much fun to miss, and despite a queue of heartbreaks that stretches several times around the block, I'm still all about falling.

Not that I've completely fallen. Nobody's dropped the 'L' word yet, but I'm definitely in lust with her. Although mixing the FIFO, my comedy and her job together will be like trying to pick a lock while blindfolded, underwater and using only a trained salmon, we've decided to have a crack, even after knowing each other barely a week. So either this is something special, or we're both idiots.

She understands that I need to work in mining to pay off debts, but doesn't know exactly how they were accumulated, and nobody needs to know everything, especially this early on – particularly the not-so-great stuff. If I made all of that first-month material, I'd be alone forever.

My plan is to try and have my first week off after only two weeks away so I can see Verity, then ask for additional time off for comedy. So it's all shimmering, wonderful goodness and positivity right now, but I'm well aware we've been existing in a magical bubble that's just about to burst, covering us both with icky obligation, real-world pressures and yucky reality.

Apart from meeting Verity at the Woodford Folk Festival, the main reason I was there was to perform. It's Australia's biggest festival of any sort, and it went well without being amazing.

One of my shows was in front of thousands, most were in front of hundreds, and a couple were in front of only tens, and it was one of these that left me properly rattled.

On the previous day, former Australian Prime Minister Bob Hawke had walked past a tent where I had hundreds in stitches. Now in his eighties, when he's not giving speeches about how Australia was far better off with him at the helm, he's puffing on a cigar, sipping a mid-strength beer, doing a sudoku and winking at all the girls. Basically living the moniker of 'iconic Australian'.

Others had told Bob good things about my act, so he made the effort to check me out, in a small bar as part of a crowd of around fifty, most of whom didn't seem to realise that a comedy performance was about to happen, and then was happening, so continued with their conversations while I attempted to put on a show.

Bob was sitting at a small table with an entourage of five, and gave me his undivided attention for ten minutes, which included not one laugh. Then he left, his five-strong posse trailing behind.

After that, I ran into Bob at Woodford a couple of times. Well, I saw him, but he didn't seem to see me.

Even after I said, 'Hello, Mr Hawke. Thanks so much for coming to my show. What did you think?'

Meaning that not only is this job on the mining site something that I need, if I'm not immediately fired, it may be something I'll have to persist with, at least until my debts are gone and I've saved a little. Possibly even enough for a mortgage, which is what so many out there are also doing.

Maybe I'll even build up enough experience that I can return to FIFO work – then if I continue to fail at the comedy and writing, I'll demote that from full-time passion to hobby. Especially if the nation decides that they agree with Bob Hawke, as they did at several elections, and stay away in their droves from my upcoming festival shows. Just like they did in Edinburgh.

Dispatch No 3 – Wednesday, January 9 (morning)

Nice nuts

Bags packed and on my way out the door, I pat my pockets to check for my phone and wallet, which I have, and keys, which I don't. Then I remember that I don't have any. I've always tried to travel as lightly as possible, but the realisation that I'm an adult with no need for keys feels a lot like failure.

At the airport, although this is my first flight anywhere for three weeks, I couldn't be less excited. Growing up, airports meant holidays. Now they're all work, either mining or comedy, and I'm also fed up with flying alone. Standing in front of the self-check-in machine struggling to remember my flight number, time and destination, I'm instead fantasising about my first holiday with Verity. Which we haven't discussed yet but I'm not getting too far ahead of anything; she's simply provided a face to my long-held daydream of travelling somewhere for fun with a partner. Something I haven't done for nearly a decade.

After twice misspelling my last name, 'Toby', I mistakenly tick that I'm carrying dangerous goods, so now need to talk to a human.

The guys I work with are among those milling around the departure gate, and I'd never forgotten that they were mega-bogans. Looking at them now for the first time in weeks, however, I realise that, such is the heights of their bogan splendour, that I'd never be able to properly commit it to memory. Designer sunglasses, so many tattoos, sparkling jewellery, gelled or shaved hair, board shorts, and tight t-shirts or singlets splashed with naked women or surfing brands, moulded around love handles and bellies, and bottomed-off by thongs. The way they're pacing around and nodding at nothing, they look like obese seagulls who've just been on a shopping spree.

There are four women out of the fifty people waiting around, and they're dressed as if for a respectable lunch at the beach with friends. Other men are there in jeans with collared shirts or some other borderline smart–casual combination, plenty are drinking beers or bourbon mixers, and anyone who didn't know better would think this was the local football team, heading off for their end-of-season trip. Not the newest members of Australia's middle class.

According to the Australian Tax Office, a middle income earner makes between $40,000 and $180,000 a year after tax, so these workers, on anywhere from $120,000 to $250,000, are technically a mix of the middle and upper classes. Not bad for a bunch of blokes who mostly never attempted university and often didn't finish high school, instead leaving early to complete a trade.

The majority of working men and women used to wear suits on planes, and I wonder if anyone ever predicted that so many with a trade or ticket instead of a degree would be so regularly flown anywhere, and paid so well.

The giveaway to everyone's true purpose is the quiet. These men haven't seen each other in three weeks, but the small pockets of chatter that have sprung up are subdued, because nobody's excited about where we're going. It's the exact opposite to the excited chatter that abounds when waiting for a flight home.

I'm in jeans and a checked shirt, with black converse – my default passable-for-public combination, and I always make an effort when flying. Maybe because airports have never completely lost that childhood sense of awe for me, and they remain places where humans are regularly launched into the sky with amazingly little ever going wrong. Or perhaps the effort is more to do with a stunning flight attendant, who last year told me that she never looks twice at most people who fly because, 'The way most of them dress these days, you see higher standards on the bus.'

While waiting for the boarding call, I browse some news websites and repeatedly check my email, Facebook and Twitter accounts, afraid that something important might come through that requires an immediate reply while I'm without reception for a couple of hours, although not once has that ever happened.

The most I share with any of my workmates is a nod of recognition and I feel guilty for not attempting a conversation, but nobody tries to talk to me.

One guy I half know approaches a guy I don't remember at all, rubs the unknown guy's stomach and I overhear him say, 'That's certainly come along over Xmas. When are you due?'

The owner of the stomach gives it a pat. 'No idea. It's already been cooking for nearly two decades.'

'Twenty years of no exercise. Quite the achievement. What do you think you'll call him? Or her?'

The owner smiles, and looks down. 'Maybe liver disease. Or diabetes?'

They both laugh, and take large swigs from their bottles of premixed bourbon and cola.

Our tickets are scanned, we walk across the tarmac, and the flight is barely two-thirds full, so I don't know why Samantha at Debitel said all the flights were booked solid. Maybe she automatically sets everyone's expectations low as a rule, so when she can't help or can't be bothered, people are less upset.

Shuffling to my seat towards the back, where I always choose to sit as it increases my odds of being next to a spare seat, I count three out of fifty people reading books. So, if these notes about my time as a FIFO are ever turned into a book, I've got at least three potential readers. Add to that the friends and family who'll actually buy the book instead of just expecting a freebie, and that's still three.

The others on the plane are at their phones, flicking through magazines of half-naked women, or staring blankly at either the safety instruction card, the inflight magazine or nothing.

The window seat next to me is empty, and there is some commotion as workers attempt to move next to their friends but are told they must sit in their assigned seats. Then the two flight attendants shuffle around the heaviest passengers, to ensure even weight distribution. If our safety can be compromised by a fat man sitting in the wrong spot, and the task of getting that right is trusted to the flight attendants, well, that's a worry.

While the fatties are being moved about some of the guys complain, they're ignored, and a massive man is directed to the empty space beside me. He sits, and his stomach takes over the armrest.

When I first began regularly travelling alone by air, I used to immediately introduce myself to anyone sitting beside me, because if something did go wrong, I couldn't think of anything worse than spending my last moments next to a stranger. Especially a hot one, because if I hadn't at least introduced myself, there'd be no chance of a quickie before death.

Once I shared this fear, minus the quickie part, with a middle-aged woman sitting beside me. The look she replied with, before returning to her book, let me know that this thought was never worth sharing again. So I no longer bother, also because I've discovered that the people who are the most talkative usually have the least to say.

Halfway through the flight, after our small and slightly stale complimentary meals have been distributed and devoured, I notice the use-by date on my 250 ml complimentary bottle of water.

'See that? I've never noticed that before,' I say to the guy with the mammoth gut next to me.

'What?'

'The use-by date. If water can go off, as a species we're all in a whole lot of trouble. Alcohol has one too, and that's used to preserve stuff.'

'So the water? Is it still okay?' asks gutsy.

'Apparently it doesn't expire until 2016.'

'So what are you on about? Are you trying to be funny or something?'

The sky remains clear for the duration of the journey and there isn't a bump, but despite having both air vents above me open, the air has become increasingly stuffy.

Then we hit the tarmac with a slap, and everyone is bounced up and out of their seats. Gutsy, next to me, had been picking from a packet of nuts, but the landing jolt caused his stomach to lurch upwards, knocking them up and out of his hands, covering us both in nuts, shells and seasoning.

I smile, raise my eyebrows at him and say, 'Aw nuts.'

'Stupid useless fucking pilot,' he replies. 'I'm not cleaning it up.'

He collects the nuts that aren't on me or the floor back into the packet, and continues eating.

After the baggage trailers are towed into the luggage cage, we get our bags straight off the trailers, and I notice I'm the only one with a rucksack. While the rest wheel away bags and sour expressions, I must look like I'm off for an adventure, and recall some of the excitement I felt when I arrived out here last year for my first swing.

Dispatch No 4 – Wednesday, January 9 (evening)

A 'cleavage' of utes

It's one of the largest coal deposits in the world, and over the ten-minute journey to the campsite we pass farmland, other accommodation villages and gates that lead to mining sites, and I'm reminded that by far the most ubiquitous landscape feature out here is nothing. In every direction, all the way to the horizon, apart from the odd something which, no matter its size, always looks inconsequential and tiny, because of all the nothing.

Untouched scrub and desert nothing, cleared and dusty expanses of nothing, idle machinery and structures where nothing's happening, all in the dull, drab and dirty nothing colours of grey, white, orange, green and yellow.

Early in the morning, as well as between 5 and 6 pm, huge bits of mining machinery, trucks, mining-spec utes and buses crowd the roads, and there is plenty of open-cut and underground coal mining occurring all around, but it is nothing compared to all the nothing.

Last year, when I first arrived, everything I knew about mining I'd gleaned from television, movies, and that theme- park roller-coaster ride where you race around in a mining cart. I've never seen anything like that out here and, even if those carts ever were part of the Australian mining industry, I'm sure they've been long decommissioned, as there's no way they'd satisfy safety standards that prohibit boots without steel caps, brisk walking and pointy hats. Extremely disappointing, as those carts would be a far more fun way to get to work than the bus.

I've since realised that the reason that nobody in the media ever reports on the nothing is that there's nothing exciting about it. At least at the beginning of this swing, I'm not eagerly looking out of the bus window for huge machinery rushing around, frequent explosions, or dwarves and dragons, as they've always been in every movie I've seen that features mining.

Instead, I look out to the horizon and, once off the bus at the campsite, I notice that the sun is setting, and the few wisps of cloud in the sky are on fire. Away in the distance are black clouds smudged to the ground, meaning rain might hit tonight, tomorrow, or miss us completely.

I walk past the neat rows of utes. There are at least forty here, and I wonder what the collective noun for them might be. A tribal-tattoo of utes? A Jägerbomb of utes? A cleavage, footy, Bali holiday, bong, spoiler, personalised numberplate, regiment or porn of utes? Maybe just a jerky of utes?

'Donga' is one of several words I've learnt since becoming a FIFO, and one of the few that isn't swearing. It's the official name for mining site accommodation, and walking towards mine down the identical rows, I'm reminded of the overwhelming jail vibe of this whole place, and I realise I've forgotten the number. So I search for my key as it's written on the tag, but already know I've left it at home. Meaning there is something more pathetic than an adult with no keys, and that's a grown man who only needs one key, but couldn't even remember that.

Luckily, the campsite manager is in. He hands me the spare and reminds me to return it during my next swing, and that we're supposed to drop off the keys before leaving anyway. I notice there's another spare for my donga still on the wall, meaning there are three copies of every key, and wonder if anyone has ever misplaced all of them, or if I did it, I'd be the first. Maybe I should just start leaving my door unlocked as so many of the other guys do.

Inside, my donga is as lifeless as I left it. Each one at this campsite is about the size of a small caravan or average bathroom, and contains a single bed, flatscreen television, wardrobe, bar fridge, and a separate shower and toilet. In inner-city Sydney, I believe they're rented out for hundreds of dollars a week and called 'luxury townhouses'.

The dongas are nothing special, but they're certainly better than backpacker accommodation, in which I've spent way too many nights as an over-thirty. A bed of nails, in my opinion, is also superior to backpacker accommodation, as a bed of nails has never given me scabies or bedbugs, or had noisy, bed-rattling sex with an American tourist on the bunk above me, then vomited all over my rucksack. Which was open.

I've always felt weird about the flatscreen TV though, which is designed to be watched from bed. Each time I've met a person with a television in their bedroom I've judged them, as bedrooms are meant for far superior activities. There was a TV in Verity's bedroom, and it's the only thing in that room that was never turned on. Sly wink.

Although the light is fading it's still oppressively hot – 41oC according to my phone – so I change out of my jeans and shirt into shorts, a singlet and thongs, then head for the dining room.

It's just after 7 pm, and there's hardly anyone smoking and drinking at the end of the rows, while before the break there were plenty, and I wonder if that's because of numerous New Year's resolutions, or if most of the others aren't back yet, although the number out and about was always a fraction of the total staying here. Similar to the rowdy passengers on public transport, it's those you notice that are the most memorable, and not the silent majority keeping to themselves.

The two small groups that have formed are carrying on the same as always, taking turns to brag about their obsessions: women, drinking, petrol-driven things, gambling and more drinking. Over the Xmas break I heard what I've since learnt is a popular term for them – 'CUBS'. It stands for 'cashed-up bogans'; however, a quick internet search reveals that outside Australia 'cubs' refers to a young 'bear', and a 'bear' is a man who's large, hairy and homosexual. It then occurs to me that in this heat, we'd all be more comfortable if we were naked, and I instantly feel weird for having those two thoughts so close together.

Right now, in my thongs, shirt and shorts, the only thing separating me from the cubs is my lack of a wedding ring and lower wages.

Night edges in slowly here if you watch for it, but look away and concentrate on something else for a moment and it'll suddenly become dark, which happens tonight over the short walk from my donga to the dining room.

After lining up with a plate and poised to start piling on food from the row of steaming bain-maries, a large woman behind that counter barks at me, 'No! Not you.'

'Pardon?'

'No singlets,' she tells me.

I glance at the men in singlets already eating, as well as those in high-vis, which is also banned at dinnertime, and say nothing else. She's likely been waiting all evening to act the enforcer, watching for someone she suspected wouldn't argue back, and we both know she's chosen wisely.

Across from my donga, two guys are chatting while sipping mid-strength beers, and notice me enter mine wearing a singlet, then leave wearing a shirt.

'Get kicked out of the dining room, did ya, mate?' asks one.

'By that big lesbian looking bird?' asks the other.

'That's right,' I reply.

'I bet you she wants to screw you,' says the other.

'A sweet looking kid like you. Be careful, she might tear it off,' adds one.

They both laugh, and I smile back. I'm guessing that I'm older than both of them, but with their deep crow's-feet and flab, they look at least five years older than me.

I head back to the dining room. I've picked a shirt with a frayed collar hoping that she'll have a go at me about it, but she doesn't even glance up while I fill my plate with meat. Verity's a vegan, so apart from dinners with my parents, I've had hardly any over the last couple of weeks, and I pile my plate high with it. I've managed to drop 10 kilos over the holidays, and will continue with the healthy eating tomorrow.

Although I get through two and a half plates of it, the meat is tough and stringy, which I mention to the two guys who don't seem to have moved from their position across from my donga.

'All the decent chefs got the arse,' one tells me.

'There's only that lesbian and a kid now,' says the other.

'This whole place is almost empty. They're moving everyone out. I think they're going to close it down,' one continues.

The other one explains that this is one of the smaller campsites, and it was only ever meant to be temporary accommodation for the construction workers as they worked on the larger campsites designed to last decades, not months.

A third guy has joined them, who I remember from my previous swings. Every night he'd be on his laptop until midnight, online gambling and drinking from an Esky, while continually offering free beers to any passers-by, who were then dragged into his monologue. Tonight he's got the Esky at his side, but no laptop, and is talking about how he's given up the horses and is committed to winning his family back.

After getting my laptop, I head back to the dining area, and set myself up in the adjoining covered and mostly empty outdoor area. Where I angle a chair and table for a view of my small tree, just as I did almost every evening last year. It doesn't seem to have changed at all over the break and, if I had to guess, I'd say it even had the same number of leaves. Then I resume working through the pile of admin around my upcoming comedy and fringe festival appearances, and although this is my first night back, I feel like I could've been sitting here for a minute or months, so little has changed.

At 10 pm I'm hopefully tired enough for sleep, so I depart for my donga, where I turn up the air conditioning, and not just because it's still oppressively hot. All that meat for dinner has done some awful things to my stomach, and whatever I eat out here usually has a similarly gassy effect. Sometimes I fantasise about how good it'd be to have a girl sharing my donga, but I'd probably explode or rupture something internally from holding in all that gas whenever she's around.

So even though it isn't great for the environment I keep the air conditioner pumping, as I fart like a champion out here so am terrified of accidentally gassing myself to death, and leaving the window open isn't an option due to the constant noise from the diesel generator and mining activity.

Mourner one: 'Such a tragedy. Poor Xavier, he had so much potential, and he was so close to paying off those debts.'

Mourner two: 'No, he wasn't actually.'

Mourner one: 'Oh. Still, he was trying. So do you know how he went? Was it suicide?'

Mourner two: 'You haven't heard? He actually farted himself to death.'

Both mourners stare at each other, trying badly to suppress giggles, which turns to laughter and catches on until everyone at the service including the celebrant is laughing loudly.

Mourner two: 'So you know what did him in? Three steaks, a full bowl of spaghetti bolognese, four sausages and three hamburgers.'

Mourner one: 'That's so gross! Eating like that, he was never going to last.'

Which is exactly what I'd just had for dinner, and I realise that if the power goes out, there's still a chance I might expire, so I also open the window a crack. Just before I find sleep, I remember that tomorrow I'll probably find out whether I'm going to get that time off or lose my job, and lay awake for over an hour, stressing about all sorts of things I can do nothing about.

Dispatch No 5 – Thursday, January 10 (morning)

Santa Claus lighting a fart

I've messed up the times, and am on the bus twenty minutes before it's due to depart for the worksite. So I sit at the back, feeling like the first to arrive at the unpopular kid's sixteenth birthday party. Other workers trickle on, and are also ticked off the driver's list. They occupy all the double seats first, starting from the front, unless they're one of the few with friends. All the remaining singles are then filled, again from the front. There are no seats left five minutes from the scheduled departure time, so the bus driver shuts the door and waits, ignoring those outside who repeatedly call out and bash on the door.

It's a fifteen-minute drive that takes around half an hour on the bus, and I've never figured out why. This morning it takes forty-five minutes as there are several long waits at sections of road that are down to one lane. The rate of upgrades is increasing since several mines are due to open soon, and for every day a deadline is missed, there are penalties and losses galore for the mining company, contractors, project managers – everyone.

Despite the air-conditioning, the air on the bus is thick and stuffy by the time we arrive, and there's an audible sigh of relief as the door opens. Not for the first time I remind myself that I haven't got it that bad – some workers from my campsite spend over three hours every day on a bus to and from their job sites.

My first site-wide morning meeting of this swing runs identically to so many from last year. There are around 200 here, and after completing a breath test, we wait. Some read the paper, the form guide, a magazine. About half stare at their phones, and the rest stare at nothing. Nearly every man then watches any time one of the dozen women come in, change seats, flick their hair or make any sound. The men's heads turn in unison, as if tugged by the same piece of string, and I notice that a few of the girls have drastically changed their hair colour and style.

Last year I noticed that no matter how late the women come in, there are always spare seats for them. I stopped trying to sit down after only one week onsite, after repeatedly being told that each empty chair wasn't actually empty.

'It looks empty,' I'd once replied.

'Sit down then, and see what happens,' I was told.

So I sat down, and nothing happened. I decided, however, that I'd prefer to start each day by standing, instead of with a threat. If I was desperate for a chair, my options were to start boxing, learn a trade, or get breast implants.

Leon, the head of onsite safety, begins this morning's meeting with a picture of a ladder propped up by a rock, followed by a broken ankle made infamous by almost weekly appearances last year, and then a Santa Claus lighting a fart with a welding-torch, which causes a murmur of mirth. Although the Santa apparently survived, we're told that he could've easily lit up his clothes or intestines, and that he's now unemployed.

'You need to be bloody careful what you put on social media and the internet and shit,' says Leon. 'Like Facebook and Twitter and Myspace and YouTube and that sort of thing.'

'What's Myspace?' asks someone.

'What's the internet?' asks someone else.

'It's like a porno mag, but for your computer and free,' says the first someone.

'There's porn on the internet?' says the second someone.

'Ask your wife. She's the one who told me about it,' says someone new.

'Sucked in, I don't have a wife.'

'That's because she's my wife now,' says another new voice.

'Shut up!' says Leon. He points at the image, 'Now Santa up there wasn't even at work, but he sent this picture around as a Xmas card to all his bosses, although it clearly shows him breaking several safety rules.'

Next Leon proudly announces that's it's been over two hundred days since the last lost-time injury (LTI) across all the worksites in the area; however, I can see three guys who I know for a fact have left work during that time due to injury. There's Sammy, who works for JRT Projects and needed several stitches after a pipe went through his face, a sparky who lit himself on fire and sustained minor burns, Damo the machine operator who dropped a pipe fitting on his big toe – as well as all those who rushed to see a doctor after Dale fed them his chilli beef jerky of death.

Several other accidents at this worksite and others have probably also been ignored to keep up that stupid streak. The general rule out here seems to be that you've done nothing wrong until you're caught, and while unspoken, it seems to be actively followed by everyone from the workers to those in charge.

After a supervisor from each contractor has mumbled about their tasks for the day, Leon quickly reads through the nine cardinal rules. He holds up a large laminated sheet and announces, 'Breaking any of these will get you instantly sacked and blacklisted, meaning you'll never work in mining again.'

The sheet clearly lists only eight rules, and there's a pause during which Leon thinks we're reading, but we're actually all waiting and wondering if anyone will speak up. If Dale were here, he would've been straight onto it.

As Leon lowers the sheet, a voice from the front says, 'There's only eight.'

'No, there's nine. I think I'd know.'

'Check the sheet, genius.'

Leon turns it around. 'Not my fault. I'm still waiting for the new graphics from HR.'

'So what is it then?' asks Donk, the JRT safety guy.

'Pardon?' Leon replies.

'The ninth rule.'

'Well, I have already said it, and you should all know it. So to repeat, it's to stay the required distance from deep holes.'

'And you shouldn't be going near any holes, unless you're wearing the proper protection,' adds Donk.

Everyone laughs, and Donk's expression doesn't change, so I've got no idea if he was trying to be funny.

'Lastly,' says Leon, 'everyone needs to be putting all waste in the correct bins. There are separate bins for paper, plastics, metal and general waste. We're really cracking down this year, because up until now, it's been pitiful. So if you're caught getting it wrong, you'll be sent into the bin to get the garbage out.'

'Even the metal bin?' asks Donk.

'That's right.'

'How safe is that? Sending a bloke into a bin that's full of metal scraps? You know, jagged edges and that.'

'It won't be a problem, if everyone put things into the right bin,' Leon replies.

We head to the carpark for stretches, and I recall some of the stories I heard over Xmas from other FIFOs about their companies going safety nuts.

One mining company makes all their employees put blocks in front of and behind their car wheels, to stop them from rolling away, and it's mandatory no matter where they're parked - even including residential streets and private driveways. One worksite has dismissed several workers for rolling up their sleeves or not properly tucking in their high-vis shirts. Another banned delivery drivers for reversing without a spotter and not wearing hard hats when outside of their trucks. Then there's the project engineer who, while onsite, nearly lost her job for moving a chair. The health and safety officer informed her that, before moving any office furniture, she was required to submit a request at least forty-eight hours in advance which explained her reasons for the move, a step-by-step procedure, and a sketch of the new office floor plan.

There's a sudden blast of dance music from a boom box, and a guy dressed identically to us in full high vis, but with the front of his head shaved and a large ponytail, is doing frenetic star jumps and yelling at everyone to hurry up and join him.

'What the hell is that?' Damo says to Donk.

'He's some mixed martial arts champion. Just started as a truck driver.'

'But what the fuck is wrong with him?'

'If he asks me, nothing, but you tell him what you like and see how that goes,' Donk replies, then stubs out his cigarette and begins doing star jumps.

Everyone shares broad grins throughout the routine, but not one person doesn't do exactly what martial-arts man demands.

Next it's the JRT Projects briefing in the smoking area. Jerome hands around the timesheet and safety sheet for everyone to sign, and this morning Jonno leads it as Dale is away.

'Is everyone fit for today's tasks? If not, you better speak up right now,' he says, between puffs on a cigarette.

There are around thirty here, although it's hard to tell exactly how many through the haze as about half are smoking. Around that same amount are obesely overweight. Despite the abundance of contradictory evidence, however, all of those I can see either nod in response to Jonno's question, or glance away.

I make sure I'm last to sign the timesheet, in order to check that everyone who was supposed to be on a flight actually arrived, and it seems that they have. Which is a relief.

Before I left for my very first swing, I was told that most guys have no trouble with their first one, as it's similar to school camp. Making new friends, no adult supervision and a bit of mischief. It's the second swing that really sorts blokes out, and apparently one in five FIFOs quit within their first few months.

Working in admin, part of my job is organising for guys to start out here, and I was surprised by how many don't make it. In my experience, about one in four who gets a job and passes the medical then never turns up for their flight out, or disappears at some other stage of the process. Which is why JRT prefers to hire those with previous FIFO experience.

Then there are those that, for whatever reason, even after they've lasted months, don't bother to resign and just never again show up, and I'd been told that was especially likely straight after the Xmas break.

So now, at ten past seven, over two hours since I woke up, Jonno has finished directing everyone to their work areas, as well as reminding them of what they're supposed to be doing, and it's finally time to start work.

Dispatch No 6 – Thursday, January 10 (afternoon)

Horny housewives

Jerome's phone is ringing for the third time in a row, and he's just staring at it. He's upgraded the Robbie Williams ringtone to the Britney Spear's classic 'Baby One More Time', which I don't mind as a song, but as a ringtone it's as awful as every ringtone.

'Mate, aren't you going to answer that?' I ask.

'Sorry.'

He switches it to silent, and it immediately starts to loudly vibrate.

'Why don't you just answer it?'

'I've already answered it twice today,' he replies. 'If I keep answering it, I won't get anything else done.'

'If you don't answer it, it's going to keep ringing.'

'I didn't think of that.' He pauses. 'It'll stop eventually.'

The phone immediately stops vibrating, and Jerome smiles at me. 'See?'

'Who was it?' I ask.

'Head office.'

Two minutes later the phone starts up again, and Jerome pauses the YouTube cat compilation he was giggling at to resume staring at it. I lean across to grab it, but he sees me coming and answers it first. I overhear that there's a problem with the timesheets, which he lets me fix rather than demanding to do it himself, so maybe there's some hope.

Once that's done, I scan my to-do list: file invoices, check all timesheets and safety logs, update drawings and plans, catalogue photos and videos, assemble claims and variations, and there are plenty of calculations and estimates I could do, but nobody has called or emailed about any of it since before the break. Also, I yawned five times while reading that list, so actually attempting any of it is a safety risk, as I might yawn so much that I strain something. Then the problem with that is, even if I sustained an injury worthy of hospital, afterwards I'd be put on 'light office duties', meaning I'd end up right back here.

So I get on with something that is urgent – comedy festival admin, and I instantly feel guilty. What we're collectively aiming to construct and then operate is mammoth, but for most, FIFO work is exactly the same as being on a production line and repeatedly performing an identical task, for over ten hours a day. Filling something up, emptying something out, driving a truck, digging a hole, oiling a machine, standing at a security checkpoint, keeping something clean, or checking and filing safety forms.

Although I'm one of the lucky few with a task that varies, I might get a raise or be considered for future work if I expend even just a minimum of effort, and even though I'm about to put my job on the line by asking for extra time off, I still can't bring myself to do any of it.

Before even beginning with my own stuff, I check news websites, my personal email and my Facebook account for anything new, but there's nothing as it's not even 8 am, and I last did all those things at 7.45 am.

There's a two-way radio on my desk so I turn it on and listen in, hoping for the same barrage of insults and jokes as last year. The irregular snippets of chatter are strictly business, however, so either everyone has lost their sense of humour over the break or left it at home. It's already over 40oC outside, and everyone's in the regulation mining uniform of a long-sleeve high-visibility shirt, long pants, steel-capped work boots, sunglasses and a hard hat, which can't be helping, but I'm sitting here in mid-twenties air-conditioned comfort and not feeling too funny either, and I'm supposed to be a comedian.

At 10 am I go outside to call a friend who's helping with my marketing and PR nearly free of charge. Twenty minutes in I turn to see Jonno watching me, his cigarette nearly done.

'Right there are ya, Xavier mate?'

'Fine thanks, Jonno. Everything okay?'

'Why are you out here? What's wrong with the air-conditioning?'

'Better reception out here,' I reply.

'Is that right?'

We both know it's a personal call, but I'm not about to admit it. Jonno then stares at me for a moment longer, shrugs and goes back into the demountable office we share with Jerome. I quickly finish up the call, notice that the phone is drenched with ear sweat, wipe it off, and then follow Jonno inside.

I was going to ask for that extra time off today, but not after that. Instead, I guiltily get to work and compare the latest version of the construction drawings to those we're working from. Sounds urgent, but we're notified of every last change, so I'm not even sure it's necessary and I think Jonno asked me to do it just to make sure I always have something to do. Laying out a pile of large drawings across my desk and getting out a red marker, however, does make me feel extremely important.

The latest plans are on a USB stick that I should've copied to my computer last year but didn't, and now it's gone from the drawer where I left it.

'Excuse me Jerome, you haven't seen a green memory stick around have you?'

'Nope.'

'Are you sure?'

He doesn't even look at me. 'I haven't touched any of your things, so please stop with your accusations, when it's clear that you're the one who's at fault.'

After emptying every one of my desk drawers, I'm surprised by the huge stacks of paper I've already accumulated after being out here barely three months – and spending less than half of that time on paid work.

It's not there, and I can't ask Jonno since he's the one that handed it to me. I'll check his office as soon as he wanders off, which is bound to happen at any moment – I wasn't even aware that he was in.

So I check the second demountable office that JRT Projects has onsite, which Donk, the safety guy, shares with Dale, the JRT onsite supervisor, and his second in charge, Ben, as well as any workers who need to photocopy, fill out paperwork, or otherwise loiter around avoiding outside.

Along each wall are shelves packed with testing equipment, safety gear and paperwork, all covered in a thick layer of dust, and the rubbish bin is overflowing and one air conditioning unit is broken, so it feels like I'm in a stinky sauna. So rather than touch anything I just ask Donk, who stops scanning for horny housewives in the area, looks me up and down as if trying to remember who I am, then shakes his head, and I escape back to the other office.

Jerome must have it and I need to go through his stuff, but in the exact opposite way that Jonno's seldom around because he's so busy, Jerome won't leave his desk until the end of the day, or someone asks him to do something.

'Donk's looking for you,' I tell him.

'He knows where I am.'

'It's his computer. He needs you to open some file for him.'

'What?'

'I don't know. He said only you could do it.'

It's as if someone has plugged Jerome into an electrical socket, he's up and out of the office so fast. Stupid people love feeling useful, because they so rarely get the chance.

The USB stick is in the first drawer of Jerome's desk that I open. A few minutes later, Jerome reappears.

'All done,' he says.

'Nice one.'

Considering I invented the whole situation, I'm curious to find out what actually happened over there. However, it's far safer not to ask.

He notices the USB stick poking out from the side of my laptop. 'What are you doing with my dongle?'

'Your what?'

He points. 'That right there. It's a USB modem, and I need it for the internet.'

'No, it's the memory stick I was looking for.'

'Are you sure about that?'

'Yes,' I reply.

'Could it be both? A memory stick and a modem?'

'No.'

'Well that explains why I couldn't get it to work.'

For a moment, I'm shocked that this is a real conversation two employed adults are actually having, and nobody's joking. Then I remember that last year Jerome tried to screw an aerial into a printer in order to connect it to the internet.

Jerome loudly bashes at his keyboard for a few minutes and then announces, 'I've just written a letter to head office about you going through my desk without my permission, which demands that you get fired. Now I've decided not to send it, this time. But if this ever happens again, I will not hesitate.'

I ignore him.

'Did you hear me?' he asks.

'Yes I did. Buddy.'

After carefully scanning all forty drawings I find a few differences, so I drop a list of them along with the marked-up drawings on Jonno's desk. He disappeared around lunchtime and hasn't been seen since. So I continue with my own work until Jonno bursts in at 4 pm and says, 'Have you finished that trenchwork estimate?'

'I could get it to you tomorrow.'

As he walks over, I frantically click away everything on my screen.

'It was supposed to be done today. Are you even working on it?'

'Just a couple more hours,' I reply.

It's the first time Jonno's mentioned that it needed to be done today and, if I argue, he'll claim that the deadline was set last year. Maybe it was, but it's not on my to-do list, and I can't even remember if I've started it, or exactly what he's talking about.

I find a spreadsheet titled 'trenchwork estimate', open it, and see a note at the top that clearly states that it was meant to be completed last year. Every task comes with a deadline though, they're always shifting and often they disappear completely. So maybe it should already be finished, but Jonno could've let me know this morning that he needed it today, and I suspect he didn't only because he discovered how urgent it was moments before he rushed in to see me. Luckily, it's already almost finished.

Jerome leaves at 4.45 pm, giving himself fifteen minutes for the two-minute walk to catch the 5 pm bus, same as every evening.

The estimate is done by 5.45 pm, and I hand the printed A3 spreadsheets to Jonno, who immediately starts scanning them.

'Is this all correct?' he asks.

'Well, it is only an estimate.'

'Don't be a smartarse. What I mean is, are there any mistakes?'

'I don't think so.'

'But how do you know? What's the total?' he asks.

'Just over $1.2 million.''The budget was $1.1 million. So that works.'

I don't know if he remembers that he already told me the budget, and what he wanted the estimate to be.

'Alright I haven't got time to look this over,' he continues, 'but I need to show it to Debitel right now. So you're definitely happy with it?'

'Sure.'

He stares at me for moment. 'Great. Now next time, get me this stuff when I ask for it, so I've got a chance to check it. This really isn't good enough, and if Peter and Scott knew it was happening, they'd go fucking nuts.'

'Okay.'

Jonno gathers together the spreadsheets, along with some drawings covered in highlighter that are nothing to do with the estimate, but I know he's going to pretend that they are, and charges outside where he stops for a smoke, then continues on to the meeting.

It's 5.55 pm, so I rush across to catch the 6 pm bus. My neck is stiff and sore, and I stretch it while waiting, causing it to click loudly. I'd forgotten how much it could hurt to move around so little for so long. Tomorrow I'll take a few walks around the site to break up the day. Still, I'm sure the guys who are stuck outside all day would trade a sore neck for their aching, sweaty and smelly everything.

At 6.05 pm there's no sign of a bus, but three others are waiting for it. There are a few utes in the carpark, so if I'm still waiting when someone appears I'll ask for a lift, or spend the night out here. These open spaces can get cold at night, but the other three men all look pretty cuddly. I give them a smile and a nod, then get out a book because I feel awkward just standing there, and then feel even more awkward as I read in front of them, instead of joining their conversation.

Dispatch No 7 – Thursday January 10 (evening)

Tradies with guns

At 6.15 pm, Ben appears in a work ute and tells me that the bus is no longer running.

'Who told you?' I ask.

'Well you're welcome to keep waiting, or you can get in.'

Already in the ute are four JRT employees. Pando and Ben in the front, with Cliff and Damo in the back, and Cliff's the smallest, so he shifts into the middle. The four of them are among the most highly regarded tradesmen with JRT, and Damo is the mammoth machine operator who probably still wants to knock me out, while Ben's the second in charge onsite after Dale.

I've got one foot in the car when Ben starts reversing, so I throw in the rest of myself, knocking off my hard hat and crashing into Cliff and Damo.

'Hang on!' I shout.

The ute skids to a stop, changes gear and leaps forward as I squash my legs into the tiny space behind the seat and pull the door shut, narrowly missing a fence post as we race past.

'Gotta get in faster if you want to keep your legs,' says Damo.

'Don't worry about your skid lid, it'll still be there tomorrow,' Pando adds.

'Why were you even wearing a hard hat?' asks Ben.

'It's the rules,' I reply.

The four of them laugh.

'Nobody gives a stuff about the rules after 5 pm. Why do you think we were back so late?' says Pando.

'Getting shit done,' Cliff adds.

One consequence of all the stringent safety is that the most dangerous jobs are done outside of regular working hours, after the safety officers have left for the day. I'd previously thought it was a rare occurrence, but over the last couple of months I've noticed something happening nearly every night I've stayed after 5 pm.

'Why didn't you get in the front? You're the tallest one here,' says Cliff.

'The front seat is taken,' Pando replies.

'I don't mind the back,' I say.

'Of course you don't,' says Damo.

'Did you do any comedy shows over the break?' Ben asks me.

'He's doing one right now, because his whole life's a joke,' says Damo.

'What are you reading?' asks Cliff.

I turn the book over in my hands. 'It's called *How Proust Can Change Your Life* by Alain de Botton. He's a philosopher, but he writes in this really accessible way.'

'Fuck reading,' says Damo.

'I read,' says Cliff.

'What, like street signs and the form guide?' asks Ben.

'I read too. The subtitles on foreign pornos,' says Damo.

'Books about war, and I read Wayne Carey's biography. Now that was interesting,' Cliff replies.

'Why's that?' asks Ben.

'He had a really fucked-up childhood,' says Cliff.

'Brendan Fevola's biography was called *In My Own Words* and he didn't even write it,' I tell them.

Nobody laughs, and Pando says, 'Why do you always interrupt Cliff?'

'And whose words was Fev supposed to use? Yours? You're such a dickhead,' says Damo.

The conversation shifts to people they all know but I don't, and it's hard to hear anything anyway, as Ben is doing 130 km/hr, even through roadworks where the speed limit is down as low as 40 km/hr, and three of the windows are open as Damo, Pando and Ben are smoking. Then Damo starts laughing and shovelling the air across to me using his hands, and I realise it's because he's farted, so I lower the only closed window.

The parts of the road that aren't being upgraded are falling apart, so when we're not going over loose patches because it's under construction, the bitumen is severely cracked and crumbling, as they were never designed for the size or number of vehicles that came with the mining industry. As we float from one side to the other, avoiding barriers, idle machinery, yellow signs and flashing lights, it never feels dangerous. Everything's blurring past so fast that none of it seems real, and it's still daylight, there's little other traffic and Ben couldn't look more relaxed, smoking with one hand, and driving and changing gears with the other.

'Hold on, boys!' says Ben.

There's a flash of grey in front of us, a huge thud, and we're airborne for a moment before crashing down with a jolt.

'You fucking destroyed that fucker!' says Pando.

'It fair dinkum exploded,' says Damo.

They're all laughing, and I spin around to see a smear of blood and fur quickly disappearing behind us, and start laughing as well.

'It was a kangaroo, right?' I ask.

'Just a small one,' says Cliff.

'You fucking obliterated it,' I reply, still laughing, while everyone else has stopped.

'What's so funny?' asks Pando.

I want to stop laughing but I can't, and I'm becoming hysterical.

'How are you a comedian? You've got a fucked sense of humour,' says Damo.

'It's properly, completely vaporised,' I say between laughs, now struggling to breathe, and with no idea what I find so funny.

'You're not supposed to swerve. Someone could've been hurt,' Ben explains.

'Yeah, we're all fine,' I say, still laughing, with tears now trailing down my cheeks.

'So what were we supposed to do, clever cunt?' asks Damo.

I grab my knees, focusing on them and my breathing, until I stop laughing.

'Maybe slow down?' I ask.

'I didn't even think of that,' says Ben.

'Maintaining your speed was the right thing to do. That's why it went under the car,' says Pando.

'I'm not sure that's a thing,' I reply.

'If you were driving, the 'roo would be fine, and we'd all be dead,' says Damo.

'I once got pulled over for hitting an animal,' says Cliff.

'But it's not illegal is it?' asks Ben.

'I swerved across double lines to hit it,' Cliff replies.

We all laugh.

'I wanted to try out my new bull bar, and it was only a fox. They're like feral right?'

'So what happened?' asks Pando.

'Well the cop had me for speeding, and crossing double lines, but there was nobody around and I wasn't doing anything unsafe. So when he came to the window and said, "Do you know why I pulled you over?" Well, I was pretty pissed off, so I said, "Because you're a tradie with a gun, and you've got stuff all else to do. Do you know mate, it takes two years to learn how to properly use a nail gun, and you got a real one after doing a twelve week fucking TAFE course?".'

We all laugh again for nearly a minute, and Ben has to slow down, he's laughing that hard.

'So he gave me the breatho, which I failed, and I lost my licence on the spot,' Cliff continues.

'Did you know that it's illegal in Victoria to leave your window open, after you've parked your car?' I ask.

'Stop talking,' says Pando.

'I heard in Victoria it's also a law that only qualified electricians are allowed to change lightbulbs,' says Cliff.

'You're a Victorian, aren't you Xavier?' asks Damo.

'That's right.'

'Of course it's fucking right,' says Pando.

'You can drink a beer in the car in New South Wales, as long as you're not driving,' says Ben.

'Good to know,' says Damo.

'Did you guys know that in Victoria, it's also illegal to wear pink hot pants after midday on a Sunday?' I ask.

'How do you know that?'

'The internet on my phone. I just did a search to see if it really is okay, like legally, to hit an animal and not stop.'

'And is it?' asks Cliff.

'Seems to be. But in Indonesia, the penalty for masturbation is decapitation. And in San Salvador, drunk drivers can be put in front of a firing squad.'

Nobody replies, and we sit in silence until we approach my campsite.

'Just stop here,' says Damo.

'I'll drop him up at the gate,' Ben replies.

'It's only like a hundred metres, and if you drop him right here, we don't have to turn around,' says Pando.

Ben continues up to the gate.

'Thanks for the lift,' I tell him.

‘You’re such a lazy fucker,’ says Damo.

‘Yeah, what makes you so special?’ asks Pando.

After I’m out of the ute, I say through the open window to Damo, ‘If I had’ve known you were going to whine about it like a little bitch, and it was going to be such an issue, I would’ve walked back.’

‘I haven’t even started, and your whole attitude stinks,’ Damo replies.

‘You stink.’

‘And you call yourself a comedian,’ says Pando.

Ben reverses away, Damo says something while looking directly at me that I don’t hear, and they all burst out laughing. They drive off to the more permanent campsite where most JRT employees are staying, only a few hundred metres away.

I’ve been back at work one day, and I already feel like I don’t know how to talk to anyone I work with, and that they all hate me. Trying to have a conversation with them is like a game that I have no idea how to win.

If I ignore them, I lose. If I’m defensive, I lose. If I’m friendly, I lose. If I retaliate, I lose.

Maybe they do hate me, or maybe it’s just how they talk. Or maybe I care so much about what they think of me only because I suspect it’s negative, and they can smell that I care, so they torment me for it.

So maybe I’m too sensitive, and it’s not them with the problem, but me.

Or maybe it’s that I was picked on in high school and it felt just like this, and I was never accepted there. So I need to be accepted here. To make amends for that six years of horror, and because I’m terrified of discovering, despite all my comedy and writing and performing and the decades since high school, that I’m still that helpless, shaking teenager who thinks he’s worthless and that so far in my life I’ve achieved nothing and gotten nowhere.

Or maybe I just need a friend out here to talk to about all this.

From now on, I’ll get the bus after work, every time I can.

After getting changed into a shirt and eating alone, I call Verity for a quick chat which lasts an hour, but I don’t mention any of this to her. Nobody wants a man who falls apart after only one night away.

I start plodding along with more comedy admin while sitting outside my donga, as I feel like drinking tonight and don’t want to be too far from my fridge. A few doors down, the ex-gambling addict has his stereo blaring rock classics. He’s crouched over his laptop, and it looks like he’s betting. As per usual, he's got a well stocked Esky beside him. So I finish my beer and put away my laptop, then wander past slowly, and accept the beer he offers.

Then he says:

'First day back and they've sacked me.

Over the holidays I turned it all around. Quit the gambling, cut down the booze, the whole bit, and they didn't even give me one full day to show 'em.

I've got debts. I owe people. And kids, and a mortgage and the rent. What am I gunna do now, eh?

I'm on casual wages too. Did you know that? Did ya?

I haven't seen nuffin' for three weeks, and they're going to pay me for a half a day today.

And that's it.

My bloody cousin did it. Not his fault, nuffin to do with him, but they made him sack his own blood, and he knows how much I need this.

He was crying more than I was.

I'm crying again. Sorry mate.

And I'm not angry, I want to be angry, I've got every right, but I'm not. Really I'm not.

I'm just sad.

I done all the right stuff. Like it was all wrong, but I fixed it, and they didn't even give me a chance.

I tried to tell 'em, but they didn't listen. They wouldn't.

My fucken eyes, ya know?

They told me it was cutbacks because the job's nearly done, but that's what they always say.

Look I get it, I do. I only got a go out here 'cos of my cuz, and last year I did miss a few days because of the drink, but this year, they didn't even give me a go.

I'm in a program thing and all. Talkin' about it, saying sorry, all sortsa shit.

It's just so fucken not fair.

My cuz, do you think he knew? We had Xmas lunch. He wouldda said something if he knew.

Three weeks off, and not a word. I'd have another job by now. Why wait three weeks?

That's like cruel and unusual ya know? Is that a legal thing? Maybe I could sue 'em.

The cunts.

It was all sorted. The ex-wife, the kids, all the money stuff. All I had to do was keep this fucken job and it was all sorted.

My ex isn't gunna want a bar of me now.

I mean, I'm gambling right now, and if you reckon that doesn't look great, well you're spot on.

But what else can I do? What else? Do you have any fucken ideas?

'Course not. 'Cos there aren't any.

I've got three grand left, and I need a lot more than that.

So I've gotta win tonight or I'm proper stuffed.

I mean I'm on the first flight out tomorrow, but what's the rush? I don't have to be anywhere.

Are you sure you don't want another beer?

No, it's no problem at all. If ya want one. Plenty left, and I'm not taking them with me.

Okay, yeah it's late. I get, it's late. I get it.

Mate, I hope you have better luck than me.'

Later that night, I overhear him having almost an identical conversation with his cousin.

The next morning he's still up, drinking and gambling. Apparently he's a little in front, but not nearly enough.

Dispatch No 8 – Friday, January 11

Those hippies, they take more drugs than anyone

An hour into the workday, Jonno comes in and gets the three folders that contain all the timesheets down from Jerome's shelves, and puts them on my desk.

'Give Xavier all the daily safety reports.'

'No,' says Jerome.

'Just do it. I can't be bothered with your shit today,' says Jonno.

Jerome stands, and theatrically drops a pen on his desk. 'No. Time is money, and I don't think this is efficient.'

'The job's nearly over, and now you decide to start thinking. Look out.' Jonno laughs. 'What I need you to do is print out every drawing. For our records.'

'Which ones?'

'All of them.'

'I've already done it,' Jerome says.

'Prove it,' says Jonno.

Jerome starts opening and closing drawers. 'I know they're here.'

'Your desk is the fucking Bermuda Triangle. Do it again, and then compare both sets. To make sure that you haven't missed any. It's extremely important.'

Jerome smiles and nods. 'I'm on it.'

Jonno tells me, 'Check all the timesheets and safety reports against what's on the system, and make a list of anywhere they don't match.'

Jonno leaves, and second in charge Ben appears. 'Jerome, I need to change the date on the flight request form I dropped off yesterday.'

Jerome absently lifts up a few pieces of paper, and then shakes his head.

'Bermuda Triangle. I told you!' Jonno calls out from just outside the door, where he's smoking.

Ben fills out a fresh form and hands it to me. 'Can you handle this?'

'That's my job,' says Jerome.

Ben stares at him. 'Shut the fuck up.'

Pando comes in with a receipt for some new work boots. 'I need to claim this.'

'Okay, I'll print you off the form,' says Jerome. 'After you've filled it out, hand it back to me. Then I enter it into the system, and in thirty days head office will let me know if it's a valid claim. Provided it is, I'll officially lodge it, and thirty days after that, if everything is approved, head office will put the money straight into your bank account.'

Pando stares at Jerome, and blinks slowly. 'Jonno told me to buy the boots, and that you'd give me the money for them from petty cash.'

'That's not how the system works,' Jerome replies.

Pando gets out a lighter, ignites the receipt and drops it on Jerome's desk, who leaps up and starts yelling, 'Fire, evacuate, fire!'

The flame starts to spread, then Ben puts it out with a glass of water.

'You've ruined everything!' says Jerome.

'What exactly is ruined?' Ben starts picking at the slightly charred papers on Jerome's desk. 'An office supplies catalogue? The instruction manual for the printer? Cruise ship brochures?'

Ben and Pando leave.

I attempt to check the hardcopy timesheets against what's on the system, but the internet is jammed as Jerome's downloading all the drawings. So I put on the kettle to make myself a coffee and while staring at it, willing it to boil, Jerome says, 'You do know there's a coffee machine in the Debitel office?'

He tells me the same thing whenever I make a coffee.

The second day back after a break is always the toughest for me, not the first. I'm so ready for that first day to suck that it's never as bad as I imagine. Then on the second day I think I'm back into it and relax, and realise I despise everything about being here.

A similar thing happens with every one of my comedy shows. Opening night is always incredibly stressful, but it's often busy so the crowd and adrenalin carry me through, even if I don't know the material as well as I should. Then that second night cuts me off at the knees with a smaller crowd, less adrenalin and fewer laughs, as I'm still memorising the material as well as trying to remember the funny stuff I said on the spot the night before. Which all leaves me questioning why I ever decided to be a comedian.

Jerome then starts blaring some awful pop-music mashup.

Turning it down only to make an hour-long personal call, right at his desk.

While continuing to jam the internet with his downloading.

Then he takes a break to chew through a seemingly endless salad with his mouth open.

His chair screeches across the floor each time he moves.

He bashes his keyboard every time anything takes more than a moment to load.

There's something up his nose that he refuses to blow out and instead, he keeps sucking it back in.

While continually asking me for things: email passwords, where to pay a parking ticket, how to resize a document just in case, 'What exactly is Twitter?', and to empty the bins because he did it just before the holidays, so it's my turn.

'Fuck this,' I say to my computer screen.

Then I stack up the folders I'm working from and my laptop, and leave.

JRT Projects has two demountable offices onsite, and I cross the small gap to the one shared between Donk the safety guy, Dale the onsite manager, and Ben the second in charge. It's also the office used by all other employees for photocopying, filling out safety forms and procrastinating.

One of the two air-conditioning units is broken, so although it's not even 11 am it's already hot, stuffy and smells like it's decomposing, as garbage from before the break has been squashed down into bins to make room instead of emptied, there's more rotten food in the fridge, and a hint of what could be dead mice in traps that were set months ago and now can't be found. JRT Projects has two apprentices onsite whose job it is to keep it clean; however, after clearing a space and placing my things on the one spare desk, I fill several garbage bags and already it smells better, but suspect I've just become used to it. While Jerome keeps the other office spotless, this one is far more heavily used and in the eight months JRT Projects has been onsite, I'm not sure it's been cleaned once.

The recycling bin I set up last year is still sitting empty in a corner, with the note on it explaining what can go in it. So I move it to a more prominent position, and Donk immediately drops in an apple core.

'That doesn't go in there,' I tell him.

'It's a bin, isn't it?'

Half an hour later I'm sweating and nodding off due to the heat, but there's no way I'd go back to Jerome.

Another coffee to spark myself up and Donk says, 'You do know there's a coffee machine in the Debitel office?'

'I do. Have you ever used it?'

'I like the instant stuff,' he replies.

'You know we put it in there for them,' says Ben.

'Cost $4,000 and we pretended it was a bathroom, charged it back to the mining company, and they signed off on it.'

'Has anyone put in a maintenance request for the air- conditioner?' I ask.

'Jerome did, months ago,' says Donk.

'Okay, I'll do another one,' I reply.

At lunchtime, any employees not smoking congregate in the crib room, a huge and garishly bright indoor space filled with plastic chairs and long tables. The whole structure appears to have been assembled out of shipping containers, meaning it's a bunch of small blocks that have been fitted together into one big block. The walls are lined with fridges, microwaves, windows covered with grates, rubbish bins, air-conditioners, and notice boards that advertise upcoming events, the days since the last incident and how to stretch. As well as being used for meals, the crib room is also where the site-wide morning meetings are held, along with every other decent- sized safety briefing or meeting.

Today the pockets of conversation are all about Debitel, the head-contractor, the six employees they've already dismissed this year, and speculation whether the cull will spread, who's safe, who's not, and why.

'Maybe things wouldn't be so bad if they'd been a bit smarter at the start of the job,' I say to Ben.

'They might've saved thousands. From what I'm hearing, they're millions in the hole,' he replies. 'Also, they might not be behind at all,' he continues. 'Only a total clusterfuck of a company loses money on a mining job. We're winding down out here, those guys were probably just no longer needed. The way rumours fly around this place, you might hear something from six different blokes, but they all heard it from the same nufty, who's as in the know as whatever he shat out that morning.'

Later that afternoon, Jerome comes into my new office. 'It's so hot in here! Was that replacement air-conditioning unit broken as well?'

'The new one never came,' says Donk.

'Yes it did.' Jerome looks around. 'Has nobody picked it up from the front gate?'

Donk and Ben stare at Jerome like he's an abstract painting that in their opinion could've be done by a five year old, and appears to have been hung on the wall upside down.

'Do you need any help with the timesheets?' Jerome asks me.

'No thank you.'

'Well if you need me, I'll be in the nice and cool office.'

Ben returns with the new air-conditioner. Followed a few minutes later by Cliff, Damo and Pando.

'That fucker Leon sent us back for more start cards,' says Pando.

'Fucking safety fuck,' Damo adds.

'I'll get you a big stack. He could've given you an official warning,' says Donk.

He opens the drawer of a filing cabinet, screams and leaps backwards, nearly falling over. Everyone but me explodes with laughter, and Donk throws a rubber snake onto the floor.

'Every time,' says Cliff.

'I've got a fucking phobia, you fuckhead,' says Donk.

'So who's going to install it?' asks Ben, nodding down at an air-conditioning unit that's slightly larger than a microwave.

'Why can't you do it?' asks Damo.

'Not qualified,' he replies.

'You built your own house from scratch,' says Pando.

'It's a safety thing,' says Ben.

'If it's a safety thing, you do need someone who's properly qualified. So it can be done safely,' says Donk.

'Is installing an air-conditioner really a safety thing?' asks Pando.

'No idea,' says Donk.

'Aren't you the safety rep?' asks Damo.

'So?' Donk replies.

The door opens, and I know everyone is hoping that it's one of the girls from the Debitel office, same as every time the door opens, but it's Leon, the Debitel head of onsite health and safety.

'I think it's hotter in here than it is outside,' he says.

'Good, you'll know,' says Ben.

'Know what?' Leon asks.

'Who can install the air-conditioner,' says Damo.

'Well, which one of you isn't fucking retarded?' Leon asks.

'Don't you need some safety thing or something?' asks Donk.

Leon glares at Donk, then points at the broken unit. 'Is that the bung one?'

We all nod.

He unplugs it, pulls it out of the wall, replaces it with the new one, and less than a minute later it's blasting chilled air.

Then Donk screams, as the snake has appeared on his desk.

'Can someone pass on a message to Peter for me?' Leon asks.

'Peter the plumber or Peter the boss?' asks Ben.

'Peter the boss. He gave me the cash, and I ordered him three grams, but then it arrived during my time off. So I just took it. Cocaine is such an amazing drug, ya know? But I feel awful, so tell him that he can have his money back, or more coke. He just has to let me know.'

'Sure,' says Ben.

'Also, good luck with the drug test tomorrow everyone,' adds Leon.

He laughs, everyone laughs along with him, and then he shoots a pair of imaginary finger pistols into the air, and leaves.

If you're going to get your drugs from anyone, I suppose it makes sense to get them from the head of site-wide health and safety. Still, it's a bit strange that the person in charge of onsite alcohol and drug testing also supplies them. Surely if there was anything to be concerned about though, nobody would be laughing. I took a few things at Woodford but nothing major, in no great quantities, and that was two weeks ago. Marijuana does take longer to get out of your system than anything else, however. So tonight, I'll ask the internet if I should be worried. Which could turn out awfully, but I'm not about to ask anyone out here for advice, and I'm pretty sure Leon wouldn't let me sit a practice test, because then everyone would want one.

The others chat about what they did over the break, while I'm too shocked to speak. Not that anyone is about to talk to me anyway.

From their stories, along with the others that I'd previously overheard, the Xmas break for most FIFO workers involved a combination of speeding, screwing, or getting smashed – and for the parents, all at the same time as recording their children's every movement on a smartphone. Which they just happen to have handy, in case you want to take a look.

'I had a huge win on the pokies, so bought an eight ball of coke and took it all over three days. Just while at home with the missus and the kids. Jeez, I got some shit done, probably knocked off a month's worth of renovations,' says Ben.

'Have a look at this.'

Cliff hands around a video of him and five friends in their underwear, throwing objects into a ceiling fan, then dodging or being hit by them as they are randomly flung around the room at high speed. Something I did while on holiday with friends, two years after high school.

'For the whole trip, one of my mates was doing coke off the toilet seat. He didn't once think to put the lid down, and nobody told him,' Cliff adds.

Now that's something else I've done, but only once.

'Oh, and my mate Sanger is looking for a job,' Cliff says to Ben.

'I thought he'd saved up two hundred grand?' says Ben.

'He did. Worked eight years straight for it too, and was going to buy a house, but then blew it in three months on hookers and cocaine. All while still living at home with his parents!'

They all laugh.

'I bought this plant that's supposed to get cats stoned,' says Pando. 'Fucking thing didn't work, but here's a video of what's supposed to happen.'

We crowd around, and watch a five-minute YouTube compilation of cats stumbling around, slipping off tables and chairs, and sleeping.

'See that?' asks Pando. 'Still land on their feet.'

'Dale came out to my property,' says Damo. 'And we spent a couple of days sinking piss, shooting shit, and blowing stuff up.'

Ned comes in, and after listening for a few minutes says, 'For Xmas I got crabs. Fucking itchy as fuck.'

He puts his hand down his pants and scratches violently.

'You haven't still got them have you?' asks Ben.

'No, but I had to shave everything, and the regrowth is itchy as.'

'I thought you were married?' asks Donk.

'I am. Happily.'

'Did you get a root?' Pando asks Donk.

'Don't tell anyone,' says Donk. He looks around, and notices how many people are in the office. 'Well, if you're going to tell anyone, make sure you just tell a few people. So I got one, and then paid for more on a whole bunch of sites, and haven't had a sniff since. I'm starting to think some of them might be broken, or maybe even a scam.'

The images on Donk's laptop suggest he hasn't given up yet.

'I got a heap of roots in Thailand,' says Pando. 'They don't make you tarp up either, if you pay a bit extra. Commando all the way!'

'What's commando?' asks Cliff.

'No dinger. Bareback,' says Pando.

'You should get yourself tested. It might drop off,' says Ben.

'Nah, you pay the extra so you get clean ones,' Pando replies.

'I paid extra,' says Ned. 'My wife got me this Asian hooker for Xmas, and we had this mad threesome. Didn't even need a babysitter, because we took turns with the kids, until it was their bedtime.'

'Why didn't you just organise for her to come over later?' I ask.

'Don't know,' says Ned. 'The other thing we did was go camping at Airlie Beach, and I got stoned and went spearfishing. Then hung the fish I caught from this street sign, and watched the ants devour them. Tiny piece by piece. It was like my very own nature documentary, ya know?'

'The Asian, did she give you the crabs?' asks Pando.

Ned laughs. 'No, my wife gave them to both of us!'

Everyone is quiet for a moment until Damo says, 'I shot a dingo and hung it from a sign, but the council took it down.'

Cliff has his phone out again. 'Have a look at this. I've been doing this thing where I find big bits of scrap metal and flatten them out. Then I mix the different bits together until it looks like something and the colours are right, and put it in a frame. What do you think?'

The guys nod and murmur their appreciation, and they're not just being nice. One piece looks like a mountainous landscape, another one resembles a classic car, and I wish I wasn't so surprised by how good they are.

'Oh, this band I'm in, we recorded a song,' says Ben. 'Want to hear it?'

'Go on,' says Pando.

'For sure,' Donk agrees.

It's a cover of *Hotel California*, and sounds decent even through the poor quality phone speaker.

'I play the drums,' he yells over the song.

'Got any gigs?' I ask, once it's finished.

'Not yet, but soon maybe. What about you? Aren't you a comedian or something?'

'I did some shows at Woodford. You know, the folk festival. Went okay.'

'Those hippies, they take more drugs than anyone,' says Ned.

'They're such fucking hypocrites,' Damo adds. 'They won't touch anything that's meat or has chemicals in it or tastes good, and then get high on shit mixed together with rat poison and horse drugs in bathtubs by bikies.'

'So many of them smoke too, with one hand,' says Pando, 'then hand you a bit of paper about how cheese causes cancer with the other.'

'Yeah. What the hell could anyone have against cheese?' Cliff agrees.

They all pause, waiting for me to mount a defence, but I already know that arguing with them is pointless, and I don't completely disagree.

At 4 pm, it's only myself and Donk in the office, and he hands me some keys.

'I've got a safety toolbox meeting. Lock up, will you?'

I know for a fact that we never lock the other office. 'Is that what always happens?'

'Well we steal stuff from the offices that aren't locked.'

'What does Jonno say about that? Does he know?'

'It was his idea,' Donk replies. 'Just stuff we need. Like fittings and safety gear. Not like computers or tools or anything.'

I'm waiting for Jonno to reappear so I can finally ask about my flights, after first showing him the timesheet and safety daily report audit that I finished hours ago. At a few minutes to five I email it to him, then rush across to the bus. It arrives right on five, and although it always slowly reverses into a proper parking area then only opens up once everyone is lined up behind the orange bunting, workers still charge at it and bang on the door before it's had a chance to properly park.

The front seats fill first, usually with those who were banging on the door, and they then proceed to sigh loudly and yell out at nobody to 'hurry up', 'close the door', and 'get fucking going for fuck's sake'. While the rest of us take a seat, and the driver always spends a few minutes ticking off names and completing the paperwork.

The process is the same, but the attitude is completely different to the mornings, as right now any delay eats into each employee's leisure time, and nobody's getting paid any extra.

Back at the campsite, I'm one of the few to thank the driver, and get in a half-hour run on the treadmill before dinner. Then I sit alone in the large patio area, my chair and table angled for a view of the only tree, and drink two mid-strength beers while discovering that even by the most conservative estimate I shouldn't test positive for any drugs unless they take a hair sample or I'm extremely unlucky.

This morning I decided that I wasn't going to phone Verity today, to give her a chance to miss me. So back in my donga, I stare at some poker championship on TV until I'm ready for sleep, but I've thought of something urgent I need to ask her. So I call and she doesn't answer, but it is nearly 10 pm, and I've already forgotten why I was calling.

Dispatch No 9 – Saturday, January 12

Just don't crash

Saturday morning and the office door is locked, and I've forgotten the keys. I'd finally gotten my hands on some important ones but couldn't look after them for even twenty-four hours, and now thirty JRT employees are milling around, waiting.

'Why the fuck did you lock the door? Nobody ever locks it,' says Ben.

'Don't worry,' says Jerome, as he opens the door to the other demountable office. 'We all make mistakes.'

JRT Projects is working across four different mining sites out here, and Jonno is in charge of all of them. Meaning that on any given day he could be anywhere, and we might all be standing out here for a while.

Jonno then materialises, opens the door and says to the group, 'What are you guys waiting for? Your work is out there, not in here.'

'Need to put our lunch in the fridge.'

'Safety forms.'

'Drawings.'

A few minutes after the doors have been opened, Ned is inside the office and stabbing at a safety form with a pencil and saying, 'Die, die, die!'

'What's wrong with you?' Ben asks him.

'It's the same stupid shit, every stupid freaking day.'

'So it should be easy.'

'It's such a waste of time,' says Ned.

'Every couple of days you have a big whinge, and what changes? So shut the fuck up, or get on the next flight and fuck off,' Ben replies.

Ned finishes filling out the form in silence.

Tradesmen are lined up at the photocopier, churning out colour copies of the areas where they're working today, and it's another part of the regular morning ritual. They run through hundreds, maybe thousands, of dollars worth of paper and ink a month – but I suppose it's nothing compared to the cost of a mistake.

'Excuse me, Jonno, have you got a moment to talk about the dates for my swing?' I ask.

'What about them?'

'I need them changed. I've emailed you about it.'

'Give me a minute,' he replies.

Ten minutes later I see his ute with him in it leave the site, so again email both bosses at head office, and once more hear nothing back.

Over the next couple of hours, there's an ongoing procession of employees through the office asking about their swings, needing help with forms, jamming the photocopier, searching slowly through the shelves and standing in front of the air-conditioner, and making it impossible for me to make any meaningful progress on my mass of comedy-related tasks. There's no danger, however, of me ever returning to Jerome. I'd rather set up my desk and chair outside in an open trench that's filled with raw sewage, crocodiles and peanut butter – just the smell of that stuff makes me gag.

'Xavier, help!' yells Donk.

He's waving his mouse around in the air, while his other hand grabs at his short, dense hair.

'What's the problem?' I ask.

He's on a website that promises 'real horny housewives in your area' and is chatting to someone who apparently just gave him an email address, but then the chat window vanished.

'Can I sit?'

'Okay,' he says.

He hovers behind me, then the web browser closes, and he tries to push me aside. 'It's gone! She's gone!'

'Give me a second.'

I'm not sure where it went, because although I like to come across as semi-computer literate, my computer knowledge really is only slightly better than the rudimentary basics known by most. The mouse cursor then flicks over a corner of the screen, the window reappears and the screen is layered with pop ups, which I click through until I uncover the email address. Then the office door opens and I rush to close browser windows filled with rows of exposed housewife bits, but Donk grabs my arm to stop me and I topple onto the floor.

It's Ben, and he tells me, 'Call Jonno. He looking for you, and why are you on the floor?'

'The porn. I was trying to shut it down.'

'Why?'

I think for a moment. The Debitel girls have already caught Donk several times, as have the JRT bosses, Jonno and the project managers from Nuscon, and nobody has ever cared.

'The email address,' Donk says to me. 'You've mucked it up.'

'You're using the same one that's on the screen? That's what she sent to you.'

'Yep. Horseandtigersharon@hornyathome.org.'

'Sorry mate, I don't think it's a real one.'

His shoulders drop. 'The root I got over Xmas wasn't a scam.'

I'd left my phone back at my desk, and there are five missed calls, all from Jonno. This is it. I'm getting fired. As I call him back, I get a little excited, despite my debts, at the prospect of being home as early as today.

'Mate, I need you to go into town to pick up some stuff. Take Ben's vehicle.'

'You do know I'm still not qualified to drive on a mining site?' I ask.

'Do we have to go through this shit again? You've already done it a bunch of times and nobody will know unless you crash. So just don't.'

In order to be legally allowed to drive on a mining site, you're required to do a short course that costs a few hundred dollars. Most FIFO workers see it as part of the overall safety rort, where the mining companies make different safety courses mandatory, and the businesses that offer those courses charge exorbitant prices, making millions from 'teaching' workers what most already know, or view as common-sense or pointless. Most businesses out here, however, will always charge as much as people are willing to pay. Also, most of the safety instructors are operating under guidelines that they know border on the ridiculous, but have been set by mining companies terrified of bad press and being sued, and the introduction of all the stringent regulations on Australian mining sites has resulted in significantly fewer injuries over time.

With all that mammoth machinery roaming around, thin unsealed roads that are always changing and drop away sharply to either side, and that mega-huge hole in the ground full of coal, driving on mining sites can be very dangerous, especially for someone like me who is about as adept at driving a manual as he is at saving for the future, sustaining a relationship and keeping track of keys.

Then if you're ever caught driving without the qualification, you are immediately fired, flown home and blacklisted, so you can never again work on any mining site anywhere in the world.

So I get the keys from Ben and rush into the car, whipping off my hard hat, squashing in and shutting the door as fast as I can, spilling notes and the contents of my backpack across both front seats, terrified that I'll take too long and they'll drive off and leave one of my legs behind, as has nearly happened several times already this year. Then I remember that I'm the one who's driving.

Driving across site I only stall twice, encounter no traffic, and security barely glances up as I pass. It takes fifteen minutes to get into town, and I properly observe the scenery for the first time this swing, as it's barely noticeable from behind the buses' darkly tinted windows, and the last time I rode anywhere in a ute I was too traumatised to appreciate anything, apart from the kangaroo we vaporised.

It might be bland out here, but it's certainly not blank. Any areas that haven't been taken over by mining remain sprawling cattle country, where quality is substituted for quantity, and the cows roam over huge arid expanses of grass, red dirt, rocks, assorted brush and the odd tree.

Only after travelling overseas have I come to appreciate this as a particularly Australian type of empty, with an overwhelming brightness, stillness and starkness, and washed- out colours that I've found nowhere else in the world. This year, depending on my mood, I've already found it calming, inspiring and terrifying, and this is still only my fourth day back.

Closer into town, I notice that the large fence that separates the mining site from the road has been toppled over during the break. I then drive through a T-intersection that's empty now, but might take ten minutes to get through at mining site peak hour, usually around 6 to 6.15 am. The sprawling 24-hour petrol station and roadhouse on the corner is packed with semi-trailers, and although I've never been inside, it apparently serves one of the biggest and unhealthiest burgers in the country. A hand-drawn sign in the window declares the travel programs it's been featured on, and plenty of FIFOs swear by it as a hangover cure.

I stick to the speed limit as the roads are tight and sometimes crumbling, and the surface is constantly changing between sealed, unsealed, partially sealed and potholed. Anytime I'm a passenger though, we're always speeding and I'm amazed there aren't more accidents. Keeping the ute from veering across the road takes constant vigilance, and trying to control any of the huge bits of mining machinery or trucks that are frequently passing is one of the most stressful jobs I can imagine. Right up there with parking cars in a multi-story lot, driving a delivery truck, or reverse parking a semi-trailer. When on foot, I struggle to make it through doorways without at least brushing the sides, and often misjudge them entirely. If I were in control of something far larger than me, that goes much faster and has a tiny margin of error, I imagine that entire cities would be in trouble.

I'm glad I'm a FIFO and not a DIDO, who drive themselves in and out from the mining site instead of being flown. Everyone has to submit travel plans before you start out here, detailing exactly how you're going to safely get to and from work, and I know from experience that these are rejected if there's even one spelling mistake. They're not enforced, however, and I doubt many DIDO workers stick to them, and instead regularly attempt to drive for hours after a long shift.

Every year fatigue fells a few, and a recent study found that accidents among the DIDOs has been on the rise. It focused on the DIDO workforce in the Bowen Basin, a coal mining area just like this one, and I still haven't mentioned where I am in Australia so it might even be this one. There's no point trying to figure it out for yourself, because I am actually somewhere in the Bowen Basin, meaning I've just told you.

As I know you were just about to put down this book and waste a bunch of time being a detective. So there you go. Now keep reading!

Anyway, this study discovered that while driving to work, up to 13 per cent of day shift workers and 23 per cent of night shift workers were falling asleep at the wheel. That's a lot, and I thought driving home would've been more dangerous. Turns out I'm right.

Many DIDO workers leave straight from work and drive three to four hours – some even drive nine hours or 900 km in a single trip. Meaning these workers have been awake for around twenty hours, which several other studies have found is the equivalent of driving when drunk.

Over the eight years leading up to 2013 almost two- thirds of all workplace fatalities involved a vehicle, and that's divided about equally between public roads and private roads, including job sites. Which all suggests that if mining companies are truly that concerned about safety, they should put a lot more effort into keeping people safe on their way to and from work, and not just while at work.

Not only do I make it into town without becoming a statistic, I don't even hit or graze anything. Then I collect the different tools and supplies as requested, and stop off at the small shopping centre for some personal supplies, including weight- loss protein powder, soy milk, tissues and moisturiser. As I'm lonely, lactose intolerant and I don't want to get fat again.

In the supermarket I roam the aisles pretending I'm back in a city, but that fantasy's ruined whenever I spot someone else in high vis. Which is unsurprisingly very easy as they're everywhere, and wearing high vis.

To exit the shopping centre parking lot, I need to put the ute into reverse for the first time that day, and I can't figure it out. This happened to me last year, and I try to remember how I did it then. Oh, that's right. I didn't. I just drove forward instead, and took out some plants and a garden bed, but in front of me this time is a concrete wall. I search the gearstick, dashboard, and around the pedals for a lever, switch or button, but find nothing. So search it all again, and twenty minutes later, after nearly snapping the gearstick and dripping with nervous sweat, I call Jonno.

'Quick question. Is there a trick to getting this thing into reverse?'

'No,' he says, and hangs up.

I try once more then phone Ben, who attempts to explain it to me, but I just can't get it, which is especially disappointing for both of us, as I'm in his ute.

So he borrows someone else's ute, drives into town and puts it into reverse for me. Turns out there isn't a trick with reverse, I just had the steering lock on.

'Let's just keep this between us,' I tell him. 'I'll buy you a six-pack.'

'Sure mate. No worries.'

On my way back to site, a truck covered in wide-load signs and flashing yellow lights approaches from the opposite direction. Behind it are two police cars, and following them are two trucks connected to each other and towing a large, double-storey weatherboard house. The afternoon wind has strengthened, and the house is barely moving but is still swerving all over the road, bending and shifting as it moves along, the curtains in the windows swaying violently, while another wide-load flashing light truck follows behind.

To avoid it, I need to get way off the road, into the mud and up against the fallen fence, and after the house has passed, I discover that I'm bogged. If I'd ever done that course which teaches you how to drive on mining sites, or anything remotely manly and car related, I imagine I'd know exactly how to deal with this. After the reversing episode, however, I'm determined to fix this myself, and am prepared to dig out the car by hand if necessary. Then I quickly figure out the four-wheel-drive system, and am back on the road in seconds.

I stop at the security gate, and the two guards are smiling widely. Together they say, 'Beep, beep, beep!' Laugh loudly, and wave me through. It's the sound of a reversing ute, and anytime I walk past anyone they repeat it. For the next five weeks.

Thanks for keeping it quiet, Ben.

Dispatch No 10 – Sunday, January 13

The fake cock

It's the day of the drug test and safety re-induction.

Both are mandatory for anyone who's been offsite for over fourteen days, and after the Xmas break, that's all of us.

I'm nervous, but in that way you feel when asked to do a random breath test, and you've only had one drink and know you're fine, but it's still a test with two possible outcomes – pass or fail.

Still, I passed the drug test to get out here without any problems, and I'd had a few tokes on a joint the week before that one, and now I'm starting to sound like a big drug taker. In the same way as a person who's been arrested or in an accident might be announced on the news as having alcohol, marijuana and ecstasy in their system, and they sound like a career criminal. However, for a lot of people I know, in their younger days that was just Friday night. Or for me, twice a year, and always around the time of these damn drugs tests.

Late last year, I was sitting alone one night at the wet mess and a news report came on about a university study which had revealed that all the stringent drug testing on mining sites didn't reduce the amount of drugs miners took. Around sixty other people were there, and they all stopped chatting, fiddling with their phones, and playing pool or darts to listen in.

Instead, according to the study, mining industry workers simply changed which drugs they took and how they took them, in order to avoid detection.

One guy said, 'Fuck! They know.'

And everyone pissed themselves laughing, because we all knew it was true, and while we hoped the mining companies didn't realise it was happening, we understood that they almost certainly did and chose to ignore it.

While some of the guys went crazy over Xmas, and I know for a fact that many hit all sorts of illegal things extremely hard during their week off, I've never seen anyone taking or dealing drugs at a mining site or campsite. However, I did hear that it happened, was told where to get it if I wanted it, and witnessed people taking herbal highs and other not-yet-illegal supposed equivalents, and then going off to work. During my time onsite last year, I also remember the odd story in the news about people at other mining sites being arrested for dealing all sorts of things, including cocaine, ice and ecstasy.

One of my jobs is mobilising new employees, which involves organising medicals along with the rest of their paperwork, and a few guys who'd successfully gotten jobs never ended up starting because they failed the piss test. Although everyone gets three chances, which is basically like saying, 'We can clearly see that you're still on it. So stop it, come back and try again, and you've got a job.'

This is a real conversation I had over the phone with one of them:

'Mate, when do I start?'

'Mate, you don't. You failed the drug test three times,' I replied.

'This is bullshit. I turned up to every single one of those tests on time, and did whatever stupid thing they wanted.'

'So?' I asked.

'Well it's bullshit, because nobody ever said I had to pass it.'

The medical reports for the guys that do end up working onsite can also be pretty terrifying, as they often say things like:

'A heavy smoker and drinker, we can't in good conscience recommend this candidate, but legally speaking, we can't not recommend him.'

'Medically, we can't stop this man from working out there, but he had severe coughing difficulties and breathing issues while filling in this form.'

'Physically he's fine, but it's worth mentioning that the words that weren't spelled correctly on this form by the candidate outnumbered those that were.'

'This applicant is morbidly obese, with early onset arthritis, breathing difficulties and a host of other medical issues. Drug tests were clear.'

'Passed all tests, but is approximately 70 kilograms overweight. That's more than I weigh.'

I wonder if anyone here today has a plan for avoiding the test. Every FIFO worker has heard stories about those who've found ways around the tests, but I've never spoken directly to someone who successfully bucked the system. Or maybe I have, and they're just clever enough to keep quiet.

The owner of a gas pipe testing company, who I shared a table with at dinner onsite one evening, did tell me about an amazing fail.

'This old school friend of mine, great bloke, bit of a battler, but his references checked out so I gave him a labouring job. I mean he'd had some problems, but had done all the steps or whatever. Then he went for the drug test, and I haven't heard from him since. I spoke to the guy who did the testing and he said he'd never seen anything like it. This bloke, my old school friend, well he must've been back on it because he was packing a fake cock and a bag of piss. Then when he went to piss in the jar but instead of squeezing the cock, which is what he was supposed to do, he squeezed the bag until the whole thing exploded. There was piss everywhere, and the fake cock hit the ground. Slap! The testing guy told me that for a second, he'd thought the bloke's dick had dropped off. He was so nervous that he'd pissed himself. With someone else's piss. Who does that?'

I glance around the crib room, looking for any signs that someone here is packing a fake cock, and the safety re-induction begins.

There's still around two hundred at this worksite, and around sixty here for this session. The safety inductions actually run weekly for anyone who's new or been away over a fortnight, so outside of the main breaks, sessions like this are always happening, just for far smaller groups.

About half the chairs are taken, and some workers are already using the tables to shield their mobile phone use from view. The lights are off, meaning the room is lit only by any natural light that has managed to struggle through the small, grate-covered windows. So although it's only 10 am, it feels like late afternoon. At one end of the crib room is a large white screen, a projector, a laptop, and Leon with his second-in-charge, Marty, who it appears will be leading today's session.

In attendance are tradesmen, onsite managers, various admin people, cleaners and labourers, but nobody any higher because although these things are mandatory, that doesn't apply if you're important enough to ignore the rules. Everyone is in the regulation long-sleeved, high-visibility shirt in yellow, orange or pink, featuring their company logo, with long trousers, sunglasses on their heads, gloves on their belts, and steel-capped work boots. Anyone based in an office can then be identified as they're usually in jeans instead of trousers, with shirts that are less stained, faded and foul-smelling.

After the ten minutes of audiovisual issues that seem to accompany any presentation out here, Marty reads the first slide, and I notice that he doesn't pause at the end of each line, but at random points in between. Also, he sometimes reads the wrong word, or the words in the wrong order, and always leaves off any word endings. For the first three slides he corrects about half his mistakes, then gives up and nobody mentions it.

Not the best person to be giving a six-hour safety induction, but at least this time around it's the only one I have to do. When I first arrived, I did five over two days – for the campsite, mining site, project management company, mining company and my company, JRT Projects. Anyone working outside of an office then also had to go away for a three-day course, as well as do the full weekend driving course.

Marty tells us that, 'One many is too injury, and zero is a, number that not acceptable.'

'Plenty of water causes the huge, risk of dehydration.'

'Lazy and human, error result in nothing getting done, even accident.'

Marty mentions the family who he's, 'Safe and healthy staying for to get home to,' and I pray that he doesn't read his children any bedtime stories. I'm an adult, and without the slides, I wouldn't have any idea what he's talking about.

I've often wondered if those who give the presentations out here receive any public speaking training, and Marty proves that they don't, or that Leon has a really sick sense of humour. He's standing off to the side, clicking through the slides for Marty when it's time, and I can't tell if he's stifling yawns or laughs.

Listening in as Marty trips over each word is quickly becoming excruciating, like watching a comic die onstage, but at least then it's usually only minutes until they're finished, and not hours. This is exactly how my friends and family must have felt when I was just starting out and sucked.

Marty seems to have a sense of how poorly he's doing, as we've each been given a worksheet, and instead of going through the questions or discussing the answers for every section, he just reads out the answers: 'A, C, B, B again...' Then when he says, 'Ask if questions any,' he doesn't look up or even pause before continuing onto the next slide.

During the morning tea break, I notice that none of the guys who bragged about taking all sorts of things over the Xmas holidays seem at all nervous about the upcoming drug test, but I can't stop worrying, despite knowing that if they're fine, then I'm double fine.

Then we're back into the safety, and at the end of the next section Marty realises he's misplaced the answer sheet. He slowly shuffles around papers looking for it, until after five minutes Leon sees that he's looking through the same pile for a third time, and tells us the answers from memory.

More sections follow, and there are now magazines, newspapers and phones out on each table, as most have given up even pretending to pay attention, tuning in only to circle the correct multiple choice answers.

After lunch, we split into groups to fill out a Job Hazard Analysis (JHA) form, and although the guys do at least one of these every day, everyone in my group claims ignorance, leaving me to write, while Ben tells me what to write down.

Then we're done, and there's a smattering of applause that gathers momentum until it's a roar, and although it started slow and sarcastic, it's now something genuine, as everyone here seems both grateful that it's over, and supportive of Marty for having a go. I imagine they'd provide a similar reaction following a torturous but earnest school Xmas pageant, where their child played a shepherd or the back end of a donkey.

Now for the drug test, and everyone's acting jovially, but I wonder how many are shitting themselves even just a little bit, like I am.

It's being done by an outside agency who now march in, two men in white coats carrying a big bucket each, and if the plan is to fill each bucket with thirty jars of urine, I hope they've got a trolley to wheel them out of here.

We're told it's going to be a spit test, meaning anyone who was planning to cheat is now carrying around a pointless fake cock and bag of foreign urine – and those are two objects I hope I'm never asked to bring to a party, because that's an event I'll be avoiding, but maybe spying on from a distance.

We're each given what looks like a white film canister, and one of the officials says, 'You've got to put the dipstick in your mouth and don't just lick it or suck it. Get it really wet, like soaking wet. Really drench that dipstick with your spit, otherwise you won't get a result, and you'll have to do the whole thing again.'

Someone shouts, 'Because nobody likes a dry dipstick.'

Then someone else mutters, 'Or to have to suck on two dipsticks in a row.'

'It's not as bad as you'd think,' says one of the four girls in the room.

Every man in the room turns to her, sees her huge grin, and laughs loudly.

The dipstick is attached to the lid, and as I remove it and am about to dip, it occurs to me that a negative result will mean a positive outcome, the same as with most medical testing, which probably leads to countless confusing conversations in very serious situations.

Any chatter has disappeared as we collectively lick, suck and spit onto our dipsticks, and while I treat mine like a lollypop, I notice various techniques in action around the room. Some move it around in their mouths like a joystick, others lick at it tentatively, as if it's a freezing cold icy-pole, a few have gobbled it whole and are sucking greedily, while others are holding the dipstick still and bobbing their heads up and down on it, searching their mouths for wetness, and poor Cliff has pushed his in too far, causing him to loudly cough and gag.

After all the dipsticks have been returned to their holders and placed on a table at the front of the room, the testing officials look them over and put around ten off to one side.

Now the test is happening I'm more worried than ever, but in a less panicked way because whatever my fate, there's nothing I can do to alter it. I wonder if anyone knew it'd be this type of test, and is wearing a fake mouth full of someone else's saliva.

'Alright, we have a problem,' official two announces.

Surely the test results can't be back that fast?

Also, there's no way they'd announce who's failed in front of the group, is there?

'If you hear your name called, please come and sit up here,' the same official says, making a show of pointing vigorously at the table that's right beside him.

He starts calling out names, and I don't hear any after he says, 'Toby Xavier.'

Those who don't know me often say my name backwards, and for ten seconds I manage to convince myself that he's talking about someone else, until he repeats it.

There are twelve of us at the front table, and the second official tells us some technical stuff about the test that I'm too worked up to hear, then hands us back our dipsticks and I've got no idea what's going on.

'Last time, we didn't get the dipsticks wet enough,' explains the person beside me.

After repeating the same suck, spit and drench process we sit back down, and this time only four names are called out, but again one of them is mine.

'Not his fault!' says Pando. 'He's never gotten anything wet before.'

'Fucking smackhead!' yells Damo.

'Too stoned to spit!' adds Cliff.

This time the officials keep us up the front after we've re-tested, and only send us away once they're satisfied. While waiting to be dismissed, I convince myself that all this is happening because positive results require far more spit and I've already failed.

An electrician asks the officials what they do to protect against tampering.

'They're either with me, or behind three locked doors,' official one replies.

'So if someone wanted to do something to the tests, they'd have to do something to you?' asks the same electrician.

'Is that a threat? Are you threatening him?' asks official two.

'Nope,' he replies. 'Asking for a friend. So if anyone's threatening you, I suppose it's him.'

Official two goes to speak but is clearly confused, so abandons whatever sentence he'd begun in his mind.

A minute later we're sent back to the main group, and official one says, 'Last problem. There's a test up here without a name on it. So I'm going to call out everyone's name, and if I don't call out your name, please let me know.'

He reads the name off each test, causing a few titters every time he mispronounces one.

'Who did I miss?'

Silence.

'If we don't work out who it is, we're going to have to repeat the test.'

There's a comically loud groan.

Official two reads out the names this time, but still nobody speaks up.

'You're grown fucking men!' he yells, then quietly adds, 'Whoever it is, you're not in any trouble.'

'Right everyone stand,' says official one. 'Then once I say your name, please sit down.'

Ten minutes later he's been through all the names twice, and everyone has sat down twice.

'We haven't got enough tests for everyone to do another one today, so we'll be back tomorrow,' says official two.

'Hey Marty,' Ben calls out to the safety guy. 'Did you do a test?'

'Of course,' he replies.

'Well I never heard your name.'

Marty pauses. 'Fuck it, you're right.'

Marty scrawls his name onto the blank canister, and everyone is too astonished to laugh. Both officials then quickly glance at each test as they drop them into their buckets, and official two announces that, 'We still have to do a proper analysis back at the lab, but these all seem negative.'

As we leave the crib room Jerome says, 'A negative result. Is that good or bad?'

'What do you think?' Pando asks him. 'Did everyone here pass, or did we fail and we're all fired?'

'I don't know, that's why I asked,' Jerome replies.

Dispatch No 11 – Monday, January 14

I'm a moron

Today I'm going to get my flights sorted, and this time I mean it.

I know this game. All my ignored emails, phone calls and sidestepped efforts to talk about it – they're trying to leave it long enough so that the answer can't be anything other than, 'Sorry, it's just too late now. You should've spoken up weeks ago.'

Well, I'm not having it.

My first comedy festival show is about four weeks away, and I'm hoping I'll be able to squeeze in a week off before that to see Verity. So what I want is to work two weeks on, one off, two on, and then two off for comedy, and back out here for four. If you're confused, I hope Jonno is as well and that it's approved before he realises exactly what's happening. Or I'm fired, and I can get the hell out of here, and I'll worry about the debt from Verity's bed. Whatever the outcome, I just want it sorted.

At morning tea, I catch Jonno's shadow in the crib room, but no Jonno. Half an hour later I overhear him in the smoking area, but nobody's there. Then his ute is parked beside our onsite office, and the next time I look, it's vanished.

Just after lunch the ute is back, so I walk in the 40°C heat from one end of the worksite to the other, and when there's no Jonno, I walk the whole thing again. Still nothing, so I retreat, and find him in the JRT office he shares with Jerome.

'Xavier. I need the number for the sand people.'

I find it for him, and he immediately hands it back to me, and asks me to organise a delivery for next week.

'Also, I need to chat about my flights today,' I say, while staring at my feet.

'Send me an email,' he replies. 'And leave those head office turkeys out of it. We stick together out here.'

'Okay.' Probably no point mentioning the dozen emails I've already sent to head office - Jonno's copied in on most of them anyway. 'You'll get back to me today, yeah?'

'We'll see.'

'It's just that there's some comedy things that I can't get out of,' I say. 'Before I started, I did make it clear that I'd need a bit of extra time off here and there.'

He looks at me like I've just told him that shit is both nutritious and delicious, and that he should try some. 'First I've ever heard about it.'

'Let me show you the dates.'

I feel like a sixteen year old girl, trying to convince her parents that a week away with her new twenty-one year old boyfriend is a great idea.

'I've got this Debitel meeting. So let's revisit this in an hour.'

Two hours later Jonno reappears, asks Donk for something, and I follow him back to his office.

'Fuck, Xavier. Mate, I had no idea you were behind me. Let me come across to you.'

At 4.30 pm, I've given up on anything getting done today. So I microwave my second lunch of a quarter chicken with potato and token greens, all covered in a gravy that, despite five minutes on high, remains a mostly gelatinous solid. I'm picking through it for anything edible when Jonno appears.

'I wouldn't feed that to my dog,' he says.

He drops the proposed itinerary onto my desk. 'This isn't going to work.'

Jonno and I both notice Donk and Damo staring at us from the other end of the office, and ignoring Donk's computer screen, where a naked man in a leather mask is chained to a brick wall and being whipped by three women dressed as schoolgirls.

'Should we go across to your office?' I ask.

'JRT Projects pays you every week, correct?' he says.

I nod.

'This job is either your number one priority,' he continues. 'Or there's no job.'

Jonno then surprises me by detailing the grand plans he has for me at JRT, and lists the opportunities there are at the different mining sites, as well as with other projects.

'Considering what you already know and everything you've picked up while out here, your next job with JRT would be second in charge to the project manager. Which would include a big pay bump. Or nothing. Your choice.'

I'm sitting, Jonno's standing. He pauses, leaning forward, eyebrows raised, waiting for an answer. At the other end of the office Donk and Dale are also waiting, while the gimp on Donk's computer is now manacled to a concrete wall, and the imitation schoolgirls are spraying him and each other with high pressure hoses.

Last year, the answer was easy. The dream comes first. Now that I know I'm still $20,000 in debt, however, that answer is a lot more complicated.

It's also astonishing to me that after spending barely half my time at work on actual work, I'm being offered this opportunity. If I stayed with JRT full time even for just a couple of years, I could really set myself up, and possibly still be able to pursue comedy and writing part-time.

Now, before making a decision, it's probably worth being honest about exactly how I got here, reflecting on where it's all likely headed, and thinking a bit about where I want to end up.

So with the original $20,000 debt, I haven't been completely truthful about how that accumulated. Sure it involved the Edinburgh Fringe Festival back in 2012, but that's only half of it..

Back in early 2012, I was working full-time as a writer and comedian, touring around Australia with a full-length show and making enough money at each comedy and fringe festival to do the next one. Everyone kept telling me, 'This is a great show. You should follow your dreams to the world's biggest arts festival, the Edinburgh Fringe.'

Which is a version of the advice given at some stage to every child ever. 'Follow your dreams, work hard, never give up and you can achieve anything.'

These days, anyone who says anything similar to me, I tell them to shut the hell up.

It takes a lot of money to get from Australia to Edinburgh. As a full-time writer and comedian though, my life savings to that point were zero dollars. So in order to get to Edinburgh, I took out three credit cards from three different banks by blatantly lying to them, which was surprisingly easy.

My plan was to break even, as I'd done at every other festival where I'd ever performed, thus paying off any debt I accumulated getting there.

The Edinburgh Fringe runs for nearly all of August. You're competing against around 3,000 shows, and I was doing two full-length shows daily, as well as various short spots around town, often performing for over three hours a day.

It all started off okay. People came to my show, and I got some good reviews, as well as some average and a couple of mean ones. Then around the half-way point I did my budgeting and calculated that, due to the huge costs involved in getting there, venue hire and existing, and my underwhelming but enthusiastic smattering of audience to that point, in order to break even, I was going to have to sell out every one of my remaining ten shows. Three times over.

So what I should've done was knuckle down and spend as little as possible, while working hard to sell as many tickets as possible to my remaining shows, in order to get out of there with the smallest debt possible.

What I did instead was ignore all my problems, and have a really good time.

Really, I've got no idea how much beer you can buy with $10,000 because I can't remember most of it, but that's exactly how much I spent.

Back in Australia I then had two options. Get a job in a city and it'd be years before I could go back to being a full- time writer and comedian. Or take a job on a mining site and pay off that debt in six months, which was an option because I've got a degree in mechanical engineering.

That's right, and I choose to attempt to make a living as a writer and comedian. Passing up a stable life for one where those who are successful are often described as 'completely unstable'.

My parents are very proud.

So the main reason I could get a job in mining was actually very little to do with engineering and a lot to do with my uncle, who works in mining.

He said, 'Do you want to work in admin, or do you want to be a labourer?'

It took me less than a second to answer, 'Admin!'

Part of the agreement was that I'd be allowed some time off for comedy, but I don't think anyone at JRT realised I'd be asking for this much. They never did ask for approximate dates when I started, and I didn't offer any.

When my attempts to talk about it last year were continually ignored, I went and booked myself into three different festivals, and was prepared to quit if I wasn't given the required time off. My spending was pretty much zero but I was being paid weekly, so I figured my debt was quickly disappearing. After paying all the festival costs and having a great time over Xmas, however, my January credit-card bills arrived, and I discovered that I was back to $20,000 in the red. Whoops.

Which means that if this job stopped now, and the Melbourne Comedy Festival, Perth Fringe and Adelaide Fringe went anywhere near as badly as Edinburgh, by the middle of this year I'd be even more in debt and again looking for a job, and would most likely have to settle for something that pays nowhere near as much as this, with far less freedom to do my own thing.

Even if I did last out here a total of six months to a year, which is how long I said I'd do it for right back when I started, there's still plenty to worry about. The shows I did over Xmas and New Year at the Woodford Folk Festival paid a total of $2,000. Not bad for a week's work, and more than I earn after tax out here at the mine, but my net profit was still zero. Just as I did in Edinburgh, again I chose having a good time over thinking long-term.

So comedy. There's no money in it, even when there's money it.

If I did decide to ditch the three festivals I've already paid for, I'd be sacrificing that money as well as anything I'd make from ticket sales, but I'd be gaining mining site wages and have no living expenses, so it'd probably work out around the same.

Right now, I feel like a moron for even considering a career in comedy.

No matter how many times I add it all up, I keep ending up with the answer I don't want.

Money is not a thing that I value. I hate it, but need it to get the things I want. Nothing outlandish, nothing ridiculous, but I do want a comfortable life and all that costs money.

Most of the men who work out here have families, which is something I definitely want for myself one day, and I'm thirty-four, so I'd better get on with it. So what I really need is a long-term partner, stability, some savings and a solid plan. What I have is none of those things and, if I pick comedy, I'd have even less of them.

I suppose I do have a partner, but I'm seldom in the same state as her, and we've known each other less than a month. So is that really a relationship that anyone would call serious?

Jonno's still hovering, waiting for an answer, and I know he expects that I'll take the job.

I'm still sitting, staring at my proposed itinerary, and I feel torn apart. I have a clear vision of what I wish my life would look like, but I suspect that

I'm the world's biggest fool for continually striving. I've had all the opportunities and education, as well as ongoing support from my parents, and I've put in thousands of hours of unpaid effort, and I'm still nowhere. Right now, to this point, all I've proven is that I can fail.

Is it even possible to be an artist and have a decent life?

Yes, of course it is.

However, is that possible for me?

Well that's a different question entirely.

If it was ever going to happen, surely it would've happened by now.

Why does it even have to be a choice? Plenty of people have solved this problem in so many different ways. Is there any harm in delaying my art for some security? I've got plenty of time – I'm probably not yet halfway through my life.

Back to Jonno, who's now been waiting about half a minute for an answer – and while that doesn't sound like long, I have changed my mind about three times every second.

'I really appreciate the offer, but I'm going to need that time off,' I tell him.

His face changes from that of a man who might lead an army, into a guy who's just asked for a phone number after a one night stand, with every intention of seeing the girl again, only to be told that she's not interested.

'So you either get that time off, or you're quitting?'

'Yeah.'

'You know,' he continues, 'if you're going to be away for extra weeks, we might need Jerome to stick around. If he wasn't here, you could move back to that other office.'

I nod along as he speaks.

'You're killing me here, Xavier mate. Do you want this job or not?'

'Of course I do,' I look down at my itinerary. 'But not without the time off.'

'Okay. Well if you're taking two weeks in a row, you'll need to stay out here until then. Can you do that?'

That'll mean working five weeks straight, and no week in between with Verity. 'Sure.'

'Well, just let head office know, and I'll leave you to sort out the rosters.'

'So that's all fine then?' I ask.

'The job's nearly done, and if you left me out here with that Jerome turkey, I'd shoot myself in the fucking head.'

After dinner that night, I call Verity to let her know and she says that it's fine, but we've never had a phone conversation that's gone for less than an hour, and this one barely lasts six minutes.

I did my best to say all the right things and we didn't argue, but I could never let her in on the complete truth. That despite the strength of my feelings for her, I'm not prepared to sacrifice this job or comedy for her. Yet.

If we're going to work, we should be able to get through this. We are, however, basing all this effort on one standout week. Which makes sense to me, as it's been years since I've met anyone I like as much as her. Although out here it's not as if I'm surrounded by options. Really, our situations couldn't be more different, and one rule that I've learnt applies equally to all areas of my life is that time kills all deals, and five weeks is a very long time.

Dispatch No 12 – Tuesday, January 15

Donut pizza

Around 8 am, Roy from Nuscon, the project managers employed by the mining company, appears in our office. As well as being the mining company's safety and contract liaison, he also leads most of the after-hours work, which occurs whenever safety needs to be ignored to get something done quickly.

Roy glances at Donk, who's repeatedly flicking a rubber band at a wall and scrolling through photos of single women apparently in our area.

'JRT is going to be at the 10 am handover inspection. Isn't that right?' he asks me.

'I think so. Let me check with Jonno and get back to you.'

I have zero idea what he's talking about.

Once he's gone, I ask Donk if he knows anything about it.

'Mate, the handover is nothing to do with me, I just know safety,' he replies.

So I call Jonno, who explains that we'll be inspecting the wet mess – including all cooking, dining and food storage areas.

'It's the part of the new campsite we're building out here that was supposed to be done before the break, and it should be a good laugh,' he says.

'Why's that?'

'Because we're the only one's who fucking finished. This inspection is supposed to verify practical completion, but the date for it was set right back at the start of the job, and there's been a few delays, so nothing's been completed, practically or otherwise.'

'So why not just reschedule it?'

'Probably because Roy can charge the mining company for two inspections if he has to do it again. Also, you're coming,' he tells me.

At 9.53 am Jonno and I arrive at the wet mess, and already there are two representatives from each of the four other contractors onsite, making for a total of ten. Everyone out here is always early, despite the common belief that most meetings are pointless. I'm often late even when I care, and that's something else I need to fix if I want to be a proper part of the adult world, even as an artist.

'Who else knows this is pointless?' Jonno asks the other contractors, who smile and laugh softly.

'When's the golf course going in?' he says to a landscaper.

'The same time as the waterslide and the whore house,' the landscaper replies.

'Well we've finished all the red lights,' says the electrician.

'That's about the only thing you have finished,' says the builder, who's with Debitel, the onsite managers.

'What about the pizza oven?' Jonno asks the kitchen contractor.

'There actually is one of those,' he replies.

'I know. Our blokes helped you hook it up,' Jonno replies.

'What about one of those tube rides?' asks the builder.

'Your guys would have to dig the hole for that, and at the pace they work, all the coal will be out of the ground before they're done,' Jonno replies.

'What's the difference between a plumber and an electrician?' says the builder.

'It's hard to tell,' he continues, before anyone can answer. 'They've both done very little but complain since they got here.'

The men then ask each other about their children, which is by far the favourite topic of conversation for any FIFO worker who's got them.

The wet mess is slightly elevated, so it offers a view of the worksite, and I notice that every structure out here meant for humans looks nearly identical, despite the size. They're all corrugated iron, metal, wood and glass, the same shape, and in green, brown and grey. Solely designed for use and then disposal and whenever I'm inside one of them, I always feel like I'm waiting to be somewhere else.

All construction here is on the edge of complete, and once finished will be mostly hidden underground or in walls and ceilings. So for all the months and $110 million dollars worth of work, the end result will be some freshly turned-over dirt, flawless grey roads, wood-chips, sporadic saplings, five larger buildings, and rows upon rows of neatly spaced and identical accommodation units, with a capacity for 1,400, that look like caravans built so cheaply that they've even come without wheels. It's as if someone took an unfinished scale model of a top–of-the-line prison, waved a wand and made it real.

Roy arrives bang on 10 am with two offsiders, a huge black American with a spherical stomach, holding a tape measure, and someone of indeterminate European origin with a carefully crafted goatee and a clipboard. Nuscon are an American company, and seem to delight in showing off their worldly credentials by dropping internationals wherever they can. Last year at the pub I met a couple of security guards they'd brought in from Texas.

Roy looks straight at me. 'I don't know who the fuck you are, but take the fucking attendance. There are fucking twenty of us here, and I want their fucking names written down at the start and end of this cunt. For the minutes and shit.'

I do a quick recount. Even with the Nuscon three, it's still only thirteen.

Jonno grabs my pen and pad. 'I know these blokes. You don't.'

After he's written down the names, he quietly tells me to make a list of any items that come up which involve JRT. 'That fella with the goatee who looks like a musketeer could be just drawing dicks for all we know.'

Roy produces a small flashlight, we all follow him into the wet mess, and his intense level of swearing continues, but I'm going to leave it out. The higher up you are, the sillier you sound when you swear, and the less idea you seem to have of how to do it. Roy drops several f-words and c-bombs into every sentence, but talks like someone from upper management addressing a shareholders meeting, so not only is his swearing unnecessary, it's also nonsensical and sounds ridiculous.

Roy says to the musketeer, 'We're just checking for a builder's clean. So we're not going to worry about every spec of dust. Write that down.'

He nods, and has been taking notes since he arrived, but hasn't spoken once.

The wet mess includes an outdoor drinking and recreation area, which isn't part of this inspection, as well as a 800-capacity dining room, kitchen, delivery bays, and walk-in fridges and freezers.

We begin in the dining area, which has been furnished with a huge close-up photo of strawberries on one wall, although I've never seen a live strawberry out here, and two potted fake trees. Roy moves at the pace of a disabled iceberg, stopping to inspect every piece of furniture, section of floor, wall and, in particular, the corners, closing in and shining his torch on anything he thinks looks suspicious, at the same time as asking the American to measure seemingly random distances and barking observations out to the contractors, although only the musketeer and I appear to be taking notes.

Jonno and I are standing at the back of the group, and he's giving a running commentary of everything Roy misses or gets wrong, quietly enough so that only I can hear.

Roy: 'There's water in this drain. This needs to be investigated.'

Jonno: 'That's what a drain is for, and he completely missed the sink above it that doesn't have any taps, and isn't connected to the drain.'

Roy: 'Right here. Look at this. Where the hell are all the tactiles?'

Jonno: 'That's not a thing. Maybe he means testicles? Is he trying to say that this building doesn't have any balls?'

Roy: 'This paintwork, I'm not sure. It seems spotty. Might need to be completely redone.'

Jonno: 'That's just dirt, which he'd know if he touched it instead of just shining that stupid torch at it.'

Roy: 'Cosmetically speaking, we need to check that these spot welds will pass a full and proper safety inspection.'

Jonno: 'Cosmetics are not something a safety inspector has ever checked, and there's no such thing as a "full and proper" safety inspection. It's just a safety inspection.'

Roy: 'This roller-door. Is it weather and cyclone proof? Needs to be checked.'

Jonno: 'He's missed that it's not wide enough for a forklift, that the bracket above it is already busted, and there's three holes in the wall off to the side, where I'm guessing someone has already backed into it.'

Roy: 'Dude, have a look at this! No tacticles here either.'

Jonno: 'Dude! Nobody's used that since *Point Break*.'

An hour in, Jonno says to Roy, 'Well that's it for us then.'

The other contractors laugh.

'I'm serious. We don't have any major work in here that hasn't already been inspected.'

'Sorry mate, you are not permitted to go anywhere. We started with twenty blokes, and that's how many we're going to finish with.'

I look around. Still only thirteen, and I wonder if the group includes seven ninjas that nobody told me about, who'll magically materialise when this is over and demand to have their names added to the list.

'These cool-room doors. Are they big enough for a forklift?' Roy asks Jonno.

'Nope. Too small.'

‘Aha! That was a JRT install, correct? You’ll have to replace it, and make the hole larger.’ He shakes his head. ‘That’s going to cost at least thirty thousand.’

‘We installed it, but Nuscon supplied the door and the instructions. So if you’d like it changed, that’s your responsibility,’ Jonno replies.

Roy takes another look at it. ‘Maybe it’ll be okay, for the intended purposes. I’ll check the specification.’ He looks at the musketeer. ‘Did you make a note of all that?’

The musketeer nods.

Roy: ‘Air conditioning condensate at kitchen entry. None provided.’

Jonno: ‘It’s on the opposite wall. He’s pointing at the electrical services access panel.’

Roy: ‘This pizza oven. We have to check these seals, especially where it connects to the roof. So someone needs to get a firehose up there, to hose the whole thing down.’

Jonno: ‘Or we could just wait until it rains.’

About two hours in, there’s still half the kitchen to go, and Jonno tells me, ‘You should get out of here.’

‘So you’re okay to take notes?’

‘I don’t know what you’ve been writing, but this idiot hasn’t said anything useful, and I can’t leave, but you can.’

‘And the attendance?’ I ask.

‘I guarantee you it won’t ever be mentioned again.’

I briefly consider telling Roy that I’m leaving, but he’s engrossed in the door seals of a portable fridge, while the American measures the distance between the fridge and the doorway, having not realised that it’s on wheels. The musketeer is still writing, and the different contractors are either shaking their heads, on their phones, looking intensely confused or whispering to each other, most likely discussing many of the same things that Jonno shared with me.

Also, I shouldn’t have mentioned my notes to Jonno. Although I’ve been writing for most of the time, they’re nothing to do with the inspection; however they do give a little insight into where my mind has been while it’s been happening.

So if you’re interested, here they are:

Check the check valve. But surely it’s already been checked? Hence the name.

Scribble scribble so bored so pointless got to stay awake scribble scribble.

How the hell am I nodding off while I'm standing up? Maybe I'm part horse? I wish.

If you're writing it looks like you're working. So keep writing.

Now tap the pad with your pen, look up like you're thinking deeply about something, and then write about it. That's the way! Really makes it look like you know what you're doing.

Let's just see then, how many letter 'a's' can I fit on one line?

aaaaaaaaaaaaaaaaaaaaaaaaaaaaaaaaaaaaaaaaaaa

43! That's got to be some sort of record.

Now to beat it.

aaaaaaaaaaaaaaaaaaaaaaaaaaaaaaaaaaaaaaaaaaaaaaaaaaa

51. Woohoo!

Maybe I should call *The Guinness Book of Records*?

I'm sure nobody else has been idiotic enough to even think of this.

However, there is no shortage of idiots in the world.

Right now, we're following one around, and he's probably on at least a quarter of a million a year.

Let's go again.

aaaaaaaaaaaaaaaaaaaaaaaaaaaaaaaaaaaaaaaaaaaaa

Only 45. Time to retire.

I peaked at 51, and it's all downhill from here.

Surely that's a good thing though? Isn't uphill supposed to be the bad one?

Counting the 'a's' probably looked very productive, so let's do that again.

Different totals. Maybe I've forgotten how to count?

Everyone thinks I'm writing down what Roy is saying. If only they knew.

I doubt they'd care.

Whenever I look up and catch someone's eye they either nod, or look worried because they're not writing anything down, and perhaps feel like they should be.

So better keep writing.

I actually think I'm yawning right now more often than I'm breathing.

Walking around slowly, so slowly and doing nothing, it's a real struggle to stay upright, awake and in a good mood. With so little to do, I focus on every little ache until it's unbearable.

Maybe it's the effect of my work boots on my lower back, because there's a slight heel there.

Heels are hellish. How do you girls do it?
I really want a pizza. Or a donut.
Donut pizza?
Now there’s a ten-million dollar idea.

Dispatch No 13 – Wednesday, January 16

Go smurf yourself

Swearing on construction sites can be pretty out of control, however, it's nothing compared to what happens on mining sites.

Imagine that a whole bunch of tradesmen were left on an island for hundreds of years, and had a chance to develop their very own language. Most words would be replaced with swear words, and that's exactly what it's like on a mining site.

They're not just insults either. Swearing can replace any word and even be used for punctuation, and sometimes it's quite poetic. I'll hear an expletive-laden tirade that's actually just a morning greeting or weekend recap and, as well as being hilarious, although I'm usually the only one laughing because to them it's just how they talk, each sentence is also extremely clear and succinct. I've even written some of them down, separated out the syllables and found that a surprising amount of their speech fits classic Shakespearean iambic pentameter, which is regarded by many as one of the finest vocal rhythms around.

There is one situation where the workers attempt to cut back on the swearing, and that's around any women onsite. Most men don't actually swear any less, they just apologise after each expletive, meaning every sentence takes around five times as long, and they sound like they've developed some sort of strange 'Lady Tourette's'.

'Good morning, fucking hot today, sorry. Hot one today, isn't it? Fucking boiling. I'm sorry, there I fucking go again. So the stupid fucking, sorry. The timesheet for fuck, damn. My bloody hours yesterday, I fucked it, stuffed it sorry, and the stupid fucking. I'm really sorry. The stupid way that head fucking office, fuck, I'm sorry, fuck! Sorry. Fuck. Doesn't matter.'

The other strategy I've heard for swearing less in front of women is to just leave all the swearing out. However, since it's come to replace most words, a lot of sentences no longer make much sense.

'Good morning...Hot today?...Boiling...My timesheet for yesterday...Wrong...Head office...Annual leave...Ya know? Doesn't matter.'

I'm not even sure why the men bother, because in my experience, the women onsite swear just as much as they do.

Also, with all the cursing flying around, swear words are so overused out here that they've lost their impact, and as a result the most offensive word is no longer an expletive. It's the word 'mate'.

It's only ever friendly if someone is using it to describe their relationship with you to someone else.

'That Xavier bloke? He's a good mate of mine.'

(Said nobody on the mining site ever.)

If used directly at you, it's never friendly. Its sits somewhere on a spectrum from sarcastic to hostile to, 'One more word and I'll punch your fucking head in. Mate.'

This year, although I've only been back out here a week, I reckon I've already heard 'mate' in direct reference to myself over a hundred times. Which isn't good.

What is wonderful though, is that at lunchtime today Jerome leaves for his week away. Those coming back will fly in tonight and start tomorrow, while those going for their week off always work half a day, which usually involves getting nothing done and repeatedly checking the clock until it's time to leave.

After Jerome departs, I immediately pack up and leave Donk, Ben, and the overflowing bins and symphony of smells, and move back to the other office, which I have to myself as Jonno isn't around.

Seconds after I'm settled, Pando explodes through the door, slams it, looks around and glares at me. Then he unloads one of the single most entertaining, poetic and extraordinary pieces of swearing I've ever heard.

It's mostly variations on the f-word or the c-word, used every which way, and dotted with the odd alternative. Problem is, in general society these words are viewed as incredibly insulting, and I'd like to give a sense of his speech without offending anyone. When performing as a comedian, it's a problem I solve by only swearing very occasionally and for emphasis, and I've also made a personal choice to never use the c-word, unless for a specific purpose.

The swearing is such a key part of Pando's rant, however, that it wouldn't work just to leave it out.

So are you familiar with the Smurfs? I am. I could tell you how Smurfette came to be, under what conditions a Smurf is born, and the age of Papa Smurf, in Smurf years. Anyway, they're known for replacing most words with the word 'smurf'. So that's what I'm going to substitute in for the swearing.

'Where the smurf is Jonno?'

'I haven't seen him, sorry,' I reply.

Pando then says to me:

'That smurfing machine out there nearly took off my smurfing smurf and now all the smurf I've been smurfing for the last smurf is completely and smurfing smurfed.

That stinking smurf of a smurf Jokka couldn't smurf a smurf into a smurf if his stupid smurfing smurf depended on it. I swear he smurfing smells like a smurf that's been smurfed, lit on fire, squashed and smurfed out of a smurf. After a week.

Tell smurfing Jonno to fire that smurf, before I smurfing smurf him into smurfland. For good.

And what the smurf do you do in here all day anyway, you lazy little smurf?

You sit in here in the air-conditioning doing smurf all. Don't you? You smurf. I swear, you're the laziest smurf to every smurf out of a smurf's smurf in all of smurf.

Actually I know what you smurfing do all day, you dirty little smurf's smurf. You just sit there in front of your smurfing laptop, looking at some dirty smurfing smurf, just smurfing yourself silly and getting smurf everywhere.

While I'm out there in the heat sweating smurf out of my smurf onto my smurf, and it's smurfed.

I've had it with this smurf. I can't wait until this smurf is over for smurf.

Once we're smurfing done with this smurfing day, I'll tell you what I'm going to do?

I'm going to smurf back to my smurfing donga at the smurfing campsite, and the first thing I'm going to do is light up a smurfing smoke and open a smurfing can.

Then I'm going to lock my smurfing door, and open up my smurfing laptop, and I'm going to put on some of the dirtiest smurf in all of smurf.

And I haven't had a decent smurf in smurf, so I'm going to smurf myself like I haven't been smurfed in smurf. And there's going to be smurf everywhere.

And my sheets are going to be like smurfing concrete. It's going to be smurfy.’

Then Pando nods as if to agree with himself and leaves, carefully clicking the door closed behind him.

Dispatch No 14 – Thursday, January 17 (morning)

A bucket of short circuits

Dale's back!

He's the JRT Projects onsite manager, reports only to Jonno, is incredibly entertaining, and appears to have lost a bunch of weight, his dirty and faded yellow high vis shirt now hanging off his still obese frame. He's also finally had a shave.

Anyone new onsite is fed his homemade chilli beef jerky, and last year it nearly sent me to hospital. Despite the terrible toll it takes on anyone who tries it, apparently there's no rule against what he's doing, so he keeps doing it.

As a person with some authority, he also frequently abuses it for no reason but his own enjoyment.

JRT Projects has two apprentices onsite and so far he has sent them off to town, the storeroom or searching around the site for:

elbow grease
sky hooks
pipe stretcher
tin of striped paint
bucket of dial tone
packet of decibels
one long weight

After that last one, the shop assistant came back empty handed and asked the apprentice, 'Is twenty minutes long enough? Or would you like to wait longer?'

He's sent Jerome into the bakery in town for a 'randee' tart, and to the Debitel office for a stack of verbal agreement forms.

Every time he sees an apprentice electrician, he asks if they've got any spare short circuits.

Just before Xmas he gave Jokka a can of spray paint and told him he needed all the air out of it, so he had to shake the can until it stopped rattling. Jokka shook it for twenty minutes, then gave it to one of the apprentices, who shook it for another ten.

If you leave your phone around and unlocked, he'll set the alarm on it for 2 am; there's ten dollars in coins superglued to the ground just outside the office he shares with Ben and Donk; and last year in the wet mess he took a nail gun, stood on Cliff's shoe, and fired into the floor next to the shoe, causing Cliff to scream, and then laugh.

At this morning's site-wide meeting in the crib room, there's the regular 200 blokes in attendance, plus nine well- watched women. After everyone's done the daily breath test there's the daily safety message from Leon, then Roy from Nuscon makes a special appearance to tell us that, 'Although this smurfing project is winding the smurf down, it's smurfing imperative that we keep smurfing focus and smurfing remain operating at full smurfing capacity, as there's still plenty of smurfing smurf to be done.'

Every morning we're given a similar supposedly urgent message. It's usually delivered by Leon, but about once a week we get Roy or some other special guest, and the messages are meant to be unique and current, but are always a variation of:

work hard, we're way behind

work harder, we're nearly there

clean up the site, it's a mess

nice work cleaning up, get back to work

work hard, keep clean, stay safe

nice work staying safe, have a barbecue, a backpack and a pat on the back

someone came to work drunk and that's bad

Following this is the JRT meeting, where the guys fill in and sign the timesheet. Every day they write down an identical number, but every day at least one of them stuffs it up, and another couple walk away without doing it at all.

These same men are trusted to work out exact angles and gradients, and make important pipework both airtight and watertight, which the vast majority of them manage without a problem. They also never forget to take a break or when and where they're meeting to drink after work, and if their pay is even one dollar out, they're straight onto it.

Billy's wife then phones me. Now I've never met Billy, but he apparently worked two weeks out here, and then disappeared. His last pay cheque was sent off months ago, and his wife is calling as according to her, it's two hours' pay short.

I phone Jonno, who says, 'We paid that useless fucker $3000 a week and he disappeared, and now she's arguing over a hundred bucks? Tell her to call head office, or ignore her. Completely up to you.'

So I ignore her. An hour later she phones back, so I give her the number for head office, who call Jonno, who then rings me.

'You're going to have to total up all of Billy's hours from the old timesheets.'

Around once a month an ex-employee or their partner rings up chasing a tiny amount of wages, and they never give in until they get all of what they believe they're owed, making the whole process usually cost way more than the total in question.

Just after lunch, I feel a presence behind me and turn to see Dale, squinting at my screen.

'Mate, you're doing way too much typing for that to be real work.'

He's got me cold – I was in the process of writing up this very diary entry.

'What are you doing? Writing a book or something?' he asks.

'Yes, actually. That's exactly what I'm doing.'

He laughs. 'You wish. As if you could ever write a book. If you were a real writer, you wouldn't be out here.'

'Really? Where should I be then?'

He laughs again. 'Not fucking here, that's for sure.'

During afternoon tea, JRT employees trickle in until the office is full of them. They all check which swing they're working, although they should already know. Over half of them ask me if they're working the last weekend in March, which is Easter, and some ask if they can change their roster, so I print them off the appropriate form.

'Just write personal reasons,' Pando tells Ben. 'Because they're not allowed to ask.'

'What about a pregnancy?' asks Ben.

'Is your wife pregnant?'

'No.'

'That only works if she's about to give birth. I've already tried it,' Pando replies.

'If you're looking for a reason to not work Easter,' I tell them, 'I think a death in the family would be more appropriate. Giving birth is more of a Xmas thing.'

They stare at me.

'Fuck you're weird,' says Pando.

'We don't want Easter off, we want to work it,' says Ben.

'Why?'

'Double time and a half for working public holidays,' Dale explains. 'Am I working Easter?'

'Doesn't look like it,' I reply.

'Write this down,' he tells me. 'Unable to fly out at that time due to requirements associated with final inspections from a storm-proofing perspective, and associated and urgent site-wide health and safety concerns. Got it?'

'Yep.'

'That's what you put on my form. If you say safety, they always approve it. That even trumps childbirth.'

For the last few of hours of the day, there's a constant procession of interruptions from people searching for Dale and Jonno. Which happens often but is particularly intense this afternoon, probably because Dale's back and Jonno hasn't been sighted.

When I first started out here, I tried to help. Then I discovered that Jonno is renowned for stating the time he'll be back, and showing up hours and sometimes days later. While Dale doesn't even specify times and is nearly always onsite somewhere, but is an expert at hiding.

Once I handed out Jonno's phone number, and a minute later he rang me.

'If someone doesn't have my number, there's a very good reason. Don't do that again.'

Next I tried taking messages, but that never worked either. Jonno and Dale always politely listened as I read them, but never followed up. Whoever left the message would then reappear, ask if I passed on the message and glare at me, certain I was lying.

I used to try harder for anyone who seemed important, but the results were always the same. So at least ten times a day, and around twenty times this afternoon, I've had this exact conversation.

'Dale or Jonno around?'

'Nope.'

'Know when they'll be back?'

'No. Sorry.'

‘Have they got their phones on them?’

‘Not sure.’

‘Are they on email?’

‘Maybe. Don’t know.’

There’s a pause, the person’s disappointment palpable, and then they leave.

It occurs to me, as I pack up for the day, that your level of importance is directly proportional to how easy you are to find. Which is why I’m always in the same place, and nobody is ever looking for me.

While bending over to put my laptop in my bag, there’s a rush of ice cold water down my back, pouring into my bum crack, and soaking my jeans. Dale roars with laughter. I didn’t even hear him come in, and Donk, Ben, Pando and Cliff are in the doorway, clapping and laughing.

I slowly stand up straight and nod. ‘You got me.’

‘Cheer up, Xavier mate,’ Dale says.

‘I’m not upset.’

‘I’m not upset,’ he repeats in an imitation of a wounded voice.

‘You’ve got off lightly,’ he continues. ‘I’ve seen guys with coins superglued to their foreheads, car horns hooked up to brake pedals, and guys trapped in portable toilets. Once, for an entire weekend.’

‘We should actually do that to him,’ says Pando, and they all laugh again.

Dispatch No 15 – Thursday, January 17 (evening)

Some guys just need a good whack

It's the RDO (rostered day off) tomorrow. So tonight, most will be drinking even more than usual.

Now, I need some friends out here. Even one will do. I've been calling Verity most nights, which is probably too much pressure on her, and she has begun only picking up my calls around half the time.

These aren't bad guys, it's just that we have completely different ways of communicating. I like talking about issues and ideas. They talk in anecdotes about gambling, sexual conquests, drink driving, fighting, blowing stuff up and crashing into things. So while all of this story I'm telling you is true, in order to fit in and maybe find a friend, over the last couple of nights I've written and rehearsed some anecdotes of my own. That way, when they start telling stories, I'll be able to join in.

We arrive at the Workers Club Hotel at 5.30 pm, I'm halfway through my first beer at 5.35 pm, but it takes Dale thirty minutes to make it to the bar. I watch him as he chats with the security guard, other contractors and some locals, and then I overhear him with the bar staff. From the questions he asks and the warmth of the responses, it's clear that these are ongoing conversations.

Sitting with Donk, Dale, Jonno, Ned and Ben, I spend about half an hour listening intently and waiting for an opening, but nobody offers me one. There are a couple of fleeting gaps where the stories I've developed could fit, but the conversation quickly moves on and they're gone. We're sitting outside as everyone but me is smoking, and I look around the other twenty-ish wooden picnic tables, and notice that among every group here, at least a couple are staring off at nothing.

I'm in a shout with Dale and Donk, and am back at the bar for my second time within the hour, meaning I'm already drunk. Another hour later, and all the conversations, including ours, have become far more animated. Watching the other groups, at times a few of them look to be on the edge of violence, but they always end with a laugh or an uncomfortable silence followed by a fresh story.

The current anecdote topic is drink driving and speeding, and they're unofficially competing for the title of drunkest and fastest, regardless of whether you got caught, crashed or made it home.

Maybe because I'm drunk and not thinking, I interrupt and say, 'Isn't it true that most car accidents are caused by either speeding or drink driving?'

Ben says, 'That only applies to people who can't drive properly. Like you.'

Dale adds, 'Tell us, Xavier mate, did you ever manage to find reverse?'

'Beep, beep, beep!' says Ned.

Half an hour later the subject is blowing stuff up, and Dale finishes up a story about the illegal fireworks he accidentally lit up all at once, which flew off in multiple directions and caused a brief human stampede, but no injuries, at the packed campsite by the beach where he spent Xmas and New Year with his family. Then I talk over Donk until everyone is listening, and explain how I once lit my hair on fire as I tried to light a joint at a house party, during my time living in Scotland.

'So how much of your hair did you lose?' asks Ned.

'I singed my eyebrows and burnt off the fringe.' (That bit did happen.)

'I'd used some gel on these bits at the side,' I continue, 'because they can stick out and look a bit shit, and that went up, and everyone started screaming, and pouring beer on my head to put it out. I was fine, but I had to get my head shaved the next day because there were all these bald patches. Lucky they tipped beer on me, and not anything stronger, hey? It even set off the smoke alarm, which wouldn't stop, so I got onto a chair and pulled it out, and took a part of the ceiling with me.' (None of that happened.)

'Sorry, why did you put gel in your hair?' asks Ben.

'I don't think gel can catch on fire,' says Jonno.

'Did you pay to fix the ceiling? I hope you at least gave them some money for the ceiling,' says Ned.

'Nobody cared,' I reply. 'We were all laughing that hard at me, sitting there with big patches of hair missing, soaked in beer and with black eyebrows.'

'Really? If it was my place, I would've kicked the shit out of you,' says Jonno.

I should be good at this. Back in the real world, at most parties I'm always one of the drunkest, most fun and last to leave. A big reason I started in comedy is because so many unfortunate and embarrassing things happen to me that if I didn't consider it material and make them into jokes, it'd all just be really depressing and pointless.

Ned, Jonno and Ben then share stories that aren't as funny as mine but get much bigger laughs, and I get it now. They talk for attention, and as friends they give each other verification through their reactions.

Being onstage and performing is easier than this. The rules for that are simple. Be funny, show no fear, and don't be mean to anyone who doesn't heckle you first. Out here it's as if they're all hecklers, but I'm no comedian. If I wasn't so desperate to fit in and be liked, I probably wouldn't want to talk to them at all. Which is exactly what any person who's struggling to fit into a group would say.

The Workers Club doesn't allow anyone in high vis to be inside after 8 pm.

'That's pretty ironic. Workers getting made to leave the Workers Club,' I say.

Donk, Ned, Ben, Jonno and Dale all stare at me blankly.

We move to the Black Nugget, the other pub in town. Which has no such rules, or carpet. Just a concrete floor that's probably hosed out once a week.

It's my shout again, and at the bar I'm served by a skimpy barmaid. They're women who wear tight and revealing lingerie, and make about as much as if they were onsite driving trucks.

'Having a good night?' she asks me.

'What's a nice girl like you doing in a place like this?' asks an overweight, silver-haired FIFO standing next to me.

'Saving for university,' she replies. 'I'm going to be a dolphin trainer.'

Carrying the drinks back, the floor is starting to sway. Again the group's at an outside table, which looks to be the exact same model as the one at the Worker's Club. Nobody else seems even slightly tipsy, and when Ben says that although he'd probably blow over the legal limit if he was pulled over by the cops, he still thinks he'd be fine to drive, and I believe him.

The story topic has moved onto violence, which I thought was rare out here because anyone involved in any physical altercation is instantly dismissed. I quickly learn, however, that unless the fight is caught on camera or there are arrests, an official report needs to be lodged for any action to be taken, and nobody here has ever heard of that happening.

Donk begins. 'Peter, and that's company director Peter, not plumber Peter.

Anyway, Peter rang me up last week and you know Sanger? That idiot who does the gas testing?

He was out at this shopping-centre job, walking along next to Big Red Ricky and doing his regular thing, whingeing and talking absolute crap.

Saying “this will be hard”, and “that’s not going to happen”. He’s one of those guys who just won’t give you a yes or no answer, ya know?

And Ricky just whacked him in the snoz. Right in the middle of the shopping centre.

It was school holidays and all, so there were kids everywhere and Ricky just popped him one, and Sanger just stood there and said nothing.

Some guys just need a good whack every now and then, ya know?’

Ben tells everyone, ‘Same thing happened to Marty. You know, Leon’s offsider?

He was at the pub, this pub, and one of the sparkies was just yapping in his ear. Yap, yap, yap. Something about safety or some bullshit.

And Marty told him, “You keep that up, and someone’s going to belt you.”

And the electrician was like, “Aw yeah. Who?”

And Marty was like, “Me. That’s who.”

The sparky thought he was joking, so kept yapping, so Marty fucking decked him.

Broke his cheek, then put on the medical report that the guy walked into a door or something, and the sparky signed it.

That electrician even had to go to hospital for some wire in his jaw or something. Sorry, I mean he was put on “light office duties” for a few days.’

Ned says, ‘The last time I bashed someone, it was this young guy. Years ago now, and I was in the dunnies at the pub having a piss, as you do.

Then this guy next to me had his pants down to his ankles, and he turns to me and he’s got this massive erection, and this big smile.

So I fucking bashed him, and his dick. Which got rid of the smile, and the boner.

Then his mates fucking appeared from somewhere, dragged me outside and gave me a good kicking.

I don’t know if they were gay or what, and I was fine after a couple of weeks, just some bruising and shit, but what the fuck? Ya know?’

Dale goes next.

‘You all hear what happened to Pando right at the start of the job, with that useless fucker?

This guy, a machine operator, he was a real dickhead. Took out five donga hookups in one day, destroying like a week's worth of work in one hit.

But there was heaps to do back then, so we had to give reasons and warnings and shit if we wanted to punt him.

So we got Pando onto him. You know, in that way when you want to get rid of someone.

We moved the guy over to the wacker packer, as not even Xavier would fuck that up, but this fella couldn't even get it started, and Pando was like, "You're so fucking useless, cunt, you can't even get it started."

All day Pando stood next to him, telling him, "You're a useless fucker, and your tickets don't mean shit. Fucking quit and fuck off. You're just fucked, you fucking useless fucker."

And the guy was like, "Don't start me. I'm a black belt in Taekwondo. I'll fucking end you."

And Pando kept it up for a couple of days, until the guy snapped and took a swing at him. Connected and all, but Pando didn't drop, and dragged him down into this trench and beat the fucking shit outta him. Like proper pummelled him.

And Peter the JRT director, he was onsite and he saw it, and he ran over from the office yelling "Fight, fight, fight!"

They only stopped when Leon came across, and the useless fucker, his head and face and ears, they were proper pulped.

Leon had wanted to fire Pando on the spot, but Peter told him that the guy had fallen into the trench, and Pando was trying to help him up, and the guy was like, "Yeah, that's exactly what happened."

You know Pando, fucking top worker. Don't want to lose him, but that useless fucker flew out the next day. Fucking black belt.

We tried the same thing with Bluey. You know, that old guy? He's been at it for years, but there are apprentices that know more than he does.

Like if you're that old and you're still doing this, there's something wrong with you, and with him, there fucking is.

But he did work out Pando was trying to get him, so now he just keeps hiding so everyone leaves him alone. Lazy fucker, but smart about it.'

Jonno then goes last.

'So Lachlan was this new bloke who was supposed to start out at one of the other mining sites a couple of days ago.

On his first night out here, before even doing one day of work, two blokes followed him home from the pub and beat the crap out of him. Put him in hospital.

Well, he had to be flown to a city hospital, and he rings me up the next day from the hospital bed, all worried about losing his job.

His Mum even called me up, all pleading and shit, telling me how much he needs this and I told her that as long as he did nothing wrong, it should be fine.

So he tells me that he stuck up for this woman, who wasn't much of a looker, but these two guys were having a go at her, and I thought, fair enough. You're trying to get your end in, fair enough.

Then it turns out that all this happened after midnight, and if you're starting work at 6 am and need to pass a breath test first thing, what are you doing at the pub after midnight before your first day?

And he's got a record. Caught his missus in bed with some bloke, so tipped lighter fluid on 'em and lit 'em up apparently. They were fine, but he did a couple of years.

So yesterday I had to fly out to the hospital where he was, all black eyes, broken arm and in a neck brace, and fire the fucker.'

I haven't been in a fight since I was sixteen, and the real ones I've seen since were brief and clumsy. So I keep drinking.

And drinking.

I don't bother sharing any more of my prepared anecdotes, because by 10 pm I can't talk.

The group splinters off, and I stumble over to Ben, who's standing with Cliff and Pando. All three are smoking, and years ago I occasionally went in for the odd drunken ciggie, so I bum one from Ben. Maybe if they see me smoking, they'll be a bit more friendly.

So I put the cigarette in my mouth, and light the wrong end.

Ben turns it around, lights the right end, and I take a big, long drag. Then proceed to vomit violently all over my shoes.

Ben and Cliff then help me onto the minibus that runs back and forth from the campsite every half an hour or so on the night before the RDO, so nobody has to drive. This isn't a service provided by the mining company, but by a FIFO worker who does it for free.

They even walk me back to my donga, where I immediately pass out.

At 2 am my alarm goes off. Bloody Dale. I go for a piss, and see in the mirror above the sink that Ben and Cliff have drawn two large cocks onto my face. Which I really hope weren't traced.

While scrubbing them off using the coarse soap that's supplied for free out here, it occurs to me that maybe most FIFOs feel at least a little like they don't fit in. The trick is to pretend, and not to agonise over it as much as I do. Really, would Ben and Cliff have drawn these cocks on my face if they genuinely hated me?

The scrubbing tears off some skin and there's a little blood, and I don't manage to completely remove the cocks, but I do reduce them to only faintly visible shadow-cocks.

Dispatch No 16 – Friday, January 18

The human skid-mark

I wake around 11 am and I feel truly disgusting. Like a skid- mark on clean underwear – completely unwanted and full of shit. My mouth feels like a furry arse, my throat's raw, my temples are throbbing, and I'm both starving and feel like I'm about to vomit. Again.

When am I going to grow up and stop doing this to myself?

Last night, I started to suspect that because so many FIFO workers are so dedicated to having a good time, that they knew something about life that I didn't. That the trick is not to bother with all this creative rubbish, to stop kidding myself that there's anything more to life and just get smashed. At no point last night, however, was I having a good time. If I hadn't been drunk, I would've been bored out of my brain. Apart from when they were telling those fighting stories – they were pretty great.

Still in bed, I google 'alcohol effects' using my phone. Every medical body in the world agrees that any more than two standard drinks a day causes lasting damage, and makes you five times more likely to get cancer. Last year, up to 6 per cent of all cancers in Australia were caused by alcohol. One big night can scar your liver, and it often doesn't show any symptoms until it's completely stuffed. Alcohol also affects your heart and makes you fat.

Drinking is actually one of the worst things you can do to your brain, according to neurologists and anyone else who's ever studied it. While the brain does grow new cells, alcohol can severely interrupt this process, and causes a long list of other horrendous brain stuff that only shows up in later life.

You need to abstain from alcohol for at least ten years before these risk factors go back down to anywhere near to non-drinker levels.

I'm thirty-four, with years of excessive drinking behind me. I can deal with a slowly deteriorating body, but I want to avoid all the serious stuff, and once my mind goes, I may as well be dead.

I silently promise that I'm going to drink a lot less from now on, and then detest myself that bit more. I've made and broken that same promise so many times, it's now meaningless.

So I can't change last night, but I am in control of today. Every RDO the mining company runs a shuttle bus into town, and I've wanted to go for a swim at the local pool ever since I found out there was one, so I hop on the midday shuttle.

A plaque at the swimming pool gate declares that it's a joint venture between the local council and the mining company, which means, guess who paid for it? The complex includes a 25 metre indoor pool, 50 metre outdoor pool, tiered seating and a cafe. Not bad for a town of under 10,000 people.

I get a lane to myself and start churning out the laps, and I'm ashamed that it's taken me this long to make it here. Although it's the RDO and around 1,200 workers have the day off, both pools are empty apart from around thirty locals and what appears to be three other FIFOs.

The locals have taken over the 25 metre indoor pool as it's warmer, and there are a few young families and groups of teenagers, as well as an underage couple who are groping each other up against a wall, and I hope they know that chlorinated water is not a form of contraception. Also, there seems to be at least three tattoos per person.

After the swim I feel much healthier, as if the water scraped last night off my skin, so I celebrate with a sausage roll and strawberry flavoured milk. Then I wander around the town centre, which takes all of seven minutes. Apart from the pool, entertainment in town is limited to a small shopping centre, two pubs and three cafes, which are packed every morning as FIFOs are mad for lattes. Right now the staff and customer numbers inside each are a two-all draw, and moving in such a leisurely way that I'm sure if anyone went any slower, they'd stop.

I extend my walk to the outskirts of town and pass a golf course, three service stations, a cricket and football oval, the creek and a playground, but most popular of all are the local KFC and McDonald's, where I'm fairly sure the employees aren't on FIFO wages. Weeks ago, before I gave up on the local chiropractor, I chatted to the receptionist who'd complained about her pay, claiming that she'd been saving up to leave town for the past three years, but wasn't even halfway there.

There are two chiropractors in town, making it the boom business as plenty of mining workers must have bung backs. After five visits my back was worse, and I remedied the problem by going instead to a city osteopath and losing weight.

Vacant shops are dotted throughout, which is supposedly due to the mining boom and a corresponding increase in temporary workers and a drop in permanent residents. There's even an abandoned cinema, and I heard from the pool lifeguard that the mining company attempted to buy and reopen it, but the owners are holding out for more than the $2.5 million they've been offered. Before the mining boom made it to this town, I imagine that for $2.5 million you could've easily bought the cinema, and had $2.4 million left over.

There may be a lack of options in town; however, I doubt the workers would make it in that often even if there weren't, considering the long hours, and the lack of weekends and enthusiasm. Back at the campsite there are darts and billiards, and the mining company puts on the odd event. Last year there was a comedy show, touch rugby tournament and golf day, and already this year we've had a snooker player who botched most of his tricks, and downed nearly one can of pre-mixed bourbon and cola between each attempt. The main leisure activity of most FIFOs, however, up there with drinking and smoking, is staring at the selection of Pay TV channels piped into every room via a flatscreen.

My stroll has now lasted over an hour, and so far I've counted twelve moving cars and spotted five other people walking around. Recently, there have been plenty of news reports about how FIFO workers are destroying regional communities, but this has been a mining town for decades. Prior to that it was all bush or farmland, and agriculture is an industry that is mostly affected by the weather and commodity prices, and I don't think FIFO farming has ever been a thing.

Maybe people have been leaving town because of a lack of opportunities, and that's an ongoing problem for regional Australia. Additionally, although there aren't any FIFO jobs available to the locals, unless they're prepared to fly out and back again, FIFO workers must be pouring millions into the local economy. Even if it does come in mostly through the two pubs, alcohol and cigarette sales, escort services and whatever gambling is available.

In some of the regions where FIFOs have come to town, there's also reportedly been an eight-fold increase in the number of sexual assaults, as well as a boom in illegal prostitution and sex trafficking. Which is a bit of a worry.

Another gripe is the lack of local services, and while there's a police and fire station, ambulance service and hospital, the closest hospital with specialist doctors is hundreds of kilometres away by either car, helicopter or plane, and the local clinic has plenty of GPs, where basic consultations go for $100 a pop.

Over Xmas I met a mental health worker based in a nearby mining town, and she told me that all the staff she knew were completely overworked. When I asked if she thought the mining companies were doing enough, she looked at me as if I'd just asked her if she'd ever met an alien.

'Are you serious? No, of course not.'

I'd have to agree with her, because of what I've seen and learnt so far in my time as a FIFO. Recently there's even been a spate of suicides among the workers – the horrible end result of a lack of services, together with other problems inherent in the industry, which all stem from working long hours and spending weeks at a time in relative social isolation, far away from personal support networks such as friends and family.

Years ago it was both the physical and mental health of workers that was constantly at risk; however, safety standards on mining sites in Australia are now far superior to what they were back then, and continue to be obsessively monitored and improved. It's as if new and changed safety rules are like superhero figurines to the mining companies, and they're so obsessed that every week they need a new one, although they've already got four warehouses full of them, all still in their original packaging.

The mental health of the workers, however, is still severely neglected. Before beginning FIFO work I was told, 'Make sure you've got an exit strategy. Guys who go out there without one just don't last.'

Never before have I had a job where someone has advised me to have a plan for when I was going to quit, even before I'd begun.

JRT rosters are three weeks on, one week off, and when I worked that last year, after being on for twenty-one days in a row, going home I felt like I was outside everything, and seven days wasn't nearly enough time to get back into it. It's like having one bite of your favourite meal, and then being forced to go back to rice cakes, and I have no idea how I'm going to last five weeks on rice cakes. I'm already sick of them, and I've barely been eating them a week.

FIFO work has to be so much harder for anyone in a long-term relationship, or with children. Verity and I have been together less than a month, and we're already struggling. Some workers hired by JRT quit after just a couple of swings, and those that give a reason nearly always say something like, 'I just can't hack it. I miss my family too damn much.'I have heard about swings where people work eight days on, six days off, and that does sound far more manageable. Even I could handle eight days of rice cakes.

While onsite, if I ever do hear a worker mention that they're missing their family or not doing so well, it's never phrased in a way that suggests they might need help or are struggling. 'Sure I miss them. But it's the way it is. Nothing anybody can do, so you just get on with it.'

Anyone who does express weakness is either ignored or dismissed, and never once did I hear a person suggest they should talk to a professional. Behind their backs and well within earshot the muttering then begins.

'He needs to just get over himself, man up and stop being a princess.'

'All that emotional shit, it's just chick stuff.'

'Mental health is only a problem if you're already mental.'

During my induction phase, I was advised that confidential mental health services are available, and I was told how to access these services. Since my induction, however, I've never once heard anyone mention these services again – at the morning meetings, the safety re-induction or anywhere.

I have heard about people who love FIFO work, and for them it's the perfect lifestyle. I've never met one though. Everyone I know considers FIFO work a temporary and unsustainable situation, where you cope with an extraordinarily challenging lifestyle in order to set up yourself or your family long-term. Sort of like a hot young person who marries an old fatty in the hope that they quickly cark it, and in that example the old fatty is definitely the Australian mining industry. Ahem, Clive, Gina, ahem.

At the Black Nugget for a late lunch of chicken parmigiana and chips, my personal hangover cure, and they aren't taking credit or debit cards, I don't have enough cash on me and the European backpacker taking orders is having trouble working his pen, so I doubt he knows the location of the nearest bank. So I approach the locals camped out at the bar, all nerves because although it's my day off, I'm in full mining-site fluro. With nobody to impress, I didn't see the point of wearing and tearing any of my own clothes.

'Excuse me, anyone know where to find the nearest ATM?' I ask.

'Which one?' three of them reply.

'The one with money in it.'

They laugh.

'No, which bank are you after?' says an old guy in a grey cardigan.

I tell them, and they talk over each other in an effort to give me the best directions.

It still takes me fifteen minutes to find it, which is a fair effort, as I earlier walked around the entire town centre in seven.

Following lunch, I'm on the 4 pm shuttle back to the campsite, and I'm still thinking about the mining companies, regional towns, and the long swings that many employed in the industry are made to work.

What the regional communities want is not FIFO workers, but for whole families to move and live out here. Which would mean the workers would get to see their families every night, and it's exactly how things were done when the big mining companies first started operating in Australia.

The only reason it's no longer done that way is because the mining companies have calculated that it's cheaper to build temporary accommodation, fly the employees out here and make them work ridiculous hours, instead of relocating entire families. They're all large corporations with shareholders, so maximising profits is the only reason they do anything. Despite all their politically correct and socially aware bluster, these companies exist for one sole reason, and it's not to make people's lives better. It's to make money.

Dispatch No 17 – Saturday, January 19 to Saturday, January 26

Don't step in any puddles

Day 11 – Saturday, January 19

Around 10 am, Jonno announces to no-one, 'We've been bloody lucky with the rain. It's supposed to be the wet season, and so far we've had nothing.'

So we all know what comes next.

Minutes later, I notice that the horizon has disappeared. An hour after that, I catch sight of the storm front, and it looks tiny. As if it's weather happening in another country.

Even when we start getting reports of flash flooding at nearby mines, the black clouds still look like someone else's problem.

At 2 pm, the black's nearly on us but is approaching from only one direction, with brilliant light-blue sky either side of it, and Jonno's confident we'll be spared.

Dale replies, 'You think?'

Donk says nothing, and keeps staring out the window.

Around 3 pm, the entire sky is dark grey, and although sunset's three hours away it's rapidly getting dark.

'Call them in,' Jonno says to Dale, who I now realise has had the two-way radio in his hand, waiting for this instruction.

Once we're in the crib room Dale says to Jonno, 'That's pretty good. We've only lost two hours of work.'

'Two hours times thirty blokes at fifty bucks an hour is still three thousand. You got that handy?'

'It's only the ones who were working outside that have stopped. Half them are still at it,' Dale replies.

He's right, and there are only about eighty in the crib room. Mostly operators and truck drivers, and anyone else who operates anything too big to fit inside, as well as their spotters and traffic managers.

'Fine,' says Jonno. 'You got fifteen hundred?'

Then it comes down, and the chatter doesn't stop, but the roar made by the rain on the aluminium roof means that everyone now has to yell to be heard, and the most frequently said words become 'what?', 'pardon?' and 'huh?'

'When do you think it's going to stop?' Jonno asks Dale.

'Huh?'

'WHEN DO YOU THINK IT'S GOING TO STOP?'

Dale scrunches up his nose. 'Not anytime soon. This is the tail end of a tropical cyclone.'

'It's the monsoon. Was bound to happen eventually,' says Jonno.

He explains that, 'Around this time of year you're always going to get a heap of rain. You even budget for it, to lose around one day out of ten and we've been fairly fortunate so far.

Last year I worked this hospital job, and it bucketed down every day for three months. Digging the trenches out there, it was like trying to make a hole in soup.

We've got a good crew here too. That hospital job had some of the laziest fuckers on earth. If they could, they'd probably get someone to fuck their own wives for them.

Every day I just did laps of that site to make sure those pricks were working. Endless laps, and they used every excuse. Toilet breaks, waiting for tools they didn't need, asking pointless questions, just wandering off for no reason.

The blokes who worked outside were even worse. They wouldn't leave that lunch room if there was even a hint of rain about, it was as if they were allergic to the stuff.

We lost so much money on that fucking job.'

Jonno looks at me. 'What the fuck are you doing here? Is it raining in the office?'

Back at the campsite that night, the rain has formed puddles everywhere and it's as if whole ecosystems have materialised, as the air is thick with so many different types of insects. After ten minutes of attempted comedy work in the outdoor area I can no longer handle the bugs feasting on my legs, so I retreat to my donga. Only the frogs seem to be happy, their loud croaking just audible above the constant rain.

There's no reception in my donga, but I want to speak to Verity, so I return to the outdoor area and we talk for over two hours. Things seem okay again, and only once the call is done do I realise that my legs and arms are blanketed in red, raised and itchy bites.

Day 12 – Sunday, January 20

The site is shut down because of the continuing torrential rain, and all the workers sit in the crib room for four hours, then are sent home and paid for eight.

Jonno isn't around but I stay anyway, and pump out some quality comedy admin.

Back at the campsite, plenty of guys have been drinking since they knocked off.

I overhear one say, 'I'm not sure there's supposed to be water around that diesel generator right there.'

'Someone should do something about that,' says another.

'I would, but I've already had a coupla cans,' says a third.

'You're not on your own there,' says the first guy.

According to most of the men out here, whenever they have a drink it's only ever a 'coupla cans', which I've learnt can mean anywhere from six to twenty.

Around 2 am, there's a mammoth bang as the generator explodes. It causes my whole donga to shake and if I had any shelves with anything on them, some of those things may've fallen onto the floor.

The power goes out, bright emergency lights come on in every donga, and the air-conditioning goes off, making sleep impossible as outside it's just under 40°C.

In the dining room, I see the accommodation manager and ask him what's happening.

'The mechanic is four hours away,' he replies.

'Isn't this campsite full of tradesmen?'

'Sure is. I could probably even fix it, but only the properly accredited fella is allowed to do it. Some insurance or safety bullshit or some fucking thing.'

Although I know everyone must be awake, and their dongas quickly getting hotter, I count only four other people wandering around, and then the rain stops. It's eerily quiet without the generator noise, but with the emergency lights on the area around the campsite is fully illuminated, and I see that the water level is just under the walkway, while to one side is a huge lake and about a third of my tree is submerged.

Day 13 – Monday, January 21

This site is still shut, the rain starting up again some hours ago. Then it stops, there's a break in the clouds, and distinct beams of sunlight come through in yellow, white and gold – it looks like it's all about to clear. Until the clouds close over, and this pattern continues for the rest of the day.

At the campsite that night the water level continues to rise, and a new generator on a raised platform has been installed. Despite the water not yet being above the level of the walkway, everything is wet and muddy, as water has been blown everywhere by intermittent gale force winds and mud has been spread by the workers, who have again been drinking since midday.

Day 14 – Tuesday, January 22

Once more, the job site is closed. In the closest capital city, it's officially the wettest week on record. In the crib room, radios everywhere are tuned in to rolling reports of which towns, suburbs and city streets are flooded, being evacuated, or under threat. All the chat is around who the FIFOs know, and who has been or could be in any way affected by the floods.

Ben gives me a lift back to my campsite as it's now too wet even for the buses. We're less driving across the site and more skidding, sliding and coasting. I sit, too terrified to move, and only when Ben is accelerating towards a pole do I yell, 'Stop!'

He swerves around it and laughs. 'Didn't even see it!'

I thought the campsite looked quiet this morning, but the carpark is now empty, and I don't see one person, pair of work boots, high-vis jacket, anything.

'Oh, I forgot,' says Ben. 'The water flooded the sewer system. Your campsite's been evacuated.'

'Pardon?'

'Grab your stuff, and I'll give you a lift to a different one. And don't step in any puddles.'

'Everyone else already knows?' I ask.

'Yep. Hurry up too, the campsite in town will fill up fast, and if you miss out there, you'll have to stay at a joint that's like a hundred ks away.'

'How would I even get there tonight?'

'No idea,' he replies.

'What about the other campsites? Are they okay?'

'Fine, but already full. It's just yours that's fucked.'

The water has submerged the walkway so I can't avoid stepping in it, but I don't see any turds floating around and there's no water in my donga.

It's dark when we arrive in town at the nearby camp, and there's a trickle of cars heading out. Once inside, I discover that plenty have already been turned away, and around fifty are waiting. After lining up for thirty minutes, registering my name and waiting over an hour, they have a room for me. I have no idea why or how, and I don't ask. Unlike my previous small tin cell, at this camp I'm on the fourth floor of a six-storey apartment block with a view of the golf course. Very swish.

Dinner is delicious, with options galore and a tick, cross or question mark above each, which indicates if it's healthy and you should be eating more, unhealthy so you should consume less, and I think the question mark is proposing a moderate amount. But to me a question mark means uncertainty, and if there's a food that those in charge are unsure about it, then maybe it shouldn't be served to anyone. Underneath each symbol is a suggested serving size, and there's even a short menu of Asian-inspired dishes. A security guard on the door then makes sure everyone swipes to get in, with cards that are programmed to only allow each person one entry during every mealtime.

Walking back to my room, I notice Debitel managers, Nuscon employees and several other important-looking people that I've never seen before, most in jeans and with greying hair and large guts. Although it's supposed to be mid- strength beers only at every campsite, those with beers are sipping from bottles of heavy without even bothering to hide them in stubby holders, which was the done thing back at my previous campsite. I'm clearly here among some genuine heavy dongas.

Day 15 – Wednesday, January 23

Breakfast's a choice of hot food, fried food, cereal, yoghurt and fruit. Or seven ticks, nine question marks, and five crosses. Judging by the plates of those around me, if this were an election, the crosses would win by a landslide.

Across at the lunch buffet, I pack containers full of cold cuts, salads, fruit and pre-packaged meals. Then spot a sign that explains there are no more lids due to the floods, as the delivery couldn't get through, and I wonder who's using container lids but not containers in such numbers that there's a shortage, and where all this food is coming from if we're cut off. Waiting in a queue at the cling wrap dispenser, I'm kept entertained by the procession of FIFOs who swear, yell and fight with it.

This morning I'm rushing, as Pando has arrived to pick me up and, although he's early, and we both know that the site will be again be closed, he's sucking impatiently on cigarettes and complaining to anyone who'll answer their phone that I'm making him late.

Before departing, I need to stop off at my room. Last year I put on over 10 kilos after just a few weeks onsite, so this time around I'm eating far less, exercising much more, and drinking a protein shake every morning and evening to help keep me feeling full, and to avoid the cravings that come from spending so long sitting down every day doing so little.

My apartment door is open while I quickly pour chocolate powder and water into my shaker cup, then give it a huge shake but the lid isn't on properly, and I spray the walls, my bed and myself with brown and yell, 'FUCK!'

At that exact moment, around ten bossy looking people walk past, stop chatting and stare. I must look like I've just shat myself, and then thrown it around for fun. Too shocked to laugh, they continue on in silence.

Well, that's one way to introduce myself to the neighbours.

I change my shirt, and Pando doesn't mention the brown on my pants, in my hair or on my face, but makes me sit on a newspaper.

Onsite it's just me and Donk in the office, while the other JRT employees are either working indoors or in the crib room, again waiting to be dismissed for the day.

He tells me, 'You're bloody lucky to be at that sweet camp. Most of the guys were sent out to that hole that's hours away in fucking woop woop. They reckon the food's terrible, the bus stinks and takes forever, and the rooms are shithouse. A lot of them even made it to the nice place before you, but got knocked back. So who'd you suck off to get a room in there?'

'I have no idea,' I reply. 'When doing that sort of thing, I'm on my knees and I don't look up, because I find that people get weird about it if you make eye contact.'

'What?' he asks.

'Don't worry.'

At dinner I remember that before I flew out for this swing, I decided to search for a genuine coal miner. So tonight, I begin my mission. Problem is, I don't really know what I'm supposed to be searching for.

I stop staring once I notice a few of the men are wearing black eyeliner. As sneakily as I can, I take a few more looks to check that I'm not imagining it. Yep, black eyeliner, and it's on more than one of them. I always wondered if there was any gay activity out here, and maybe that's the sign you're up for it. If not, what else could it be? Most of the men wearing eyeliner are also in short shorts and singlets.

Day 16 – Thursday, January 24

The sky is clear and work onsite has resumed, both inside and out; however, my phone isn't getting any reception. I go outside and wave it around in the air, trying to pick up a signal, and notice three guys doing the same thing.

'Stop that, you look like a dickhead,' Dale tells me. 'Floodwaters have knocked out the mobile towers. Half the state is without coverage.'

'How do you know that?' I ask.

'Radio,' he replies. 'For a smart fucker, sometimes you're dumb as shit.'

Most of the men are in an awful mood all day, as last night the shops in town lost network access and their registers and card facilities stopped working, so they haven't been able to sell anyone anything, including cigarettes. Those without any are continuingly badgering those with any left, until just about everyone is constantly bickering.

In the early afternoon I guess the shops are back online, as the smoking area suddenly becomes standing room only.

Just before I leave for the day Jonno comes in, and I realise I haven't seen him all week.

'Each site we're working at, once we're done, we need to provide a user manual for all the stuff we've installed. So have a look through the specification and make a list of everything we've put in. Then find the user manual for each different part and just invent any bits that are missing. Then whack 'em together and write something that links it all up or something, so it doesn't sound too shit, and that's it. We're working across four sites, and each manual is about a hundred pages, but it's mostly a cut and paste job. Think you can handle that?'

I'm not entirely sure what he's talking about, but I did similar things when I was an engineer. 'Shouldn't be too hard. Can you email me some examples, from similar jobs?'

With this type of task, someone has always done something very similar before.

‘Will do, chief. Really appreciate it.’

Day 17 – Friday, January 25

My phone regains reception around 4 pm. No messages.

That night, I quietly ask the security guard on the door of the dining room, ‘What’s the deal with the eyeliner?’

‘The what?’

He’s probably as embarrassed as I feel. I repeat, even more quietly. ‘The black stuff, around some of the blokes’ eyes.’

‘Oh, that’s coal. They’re coal miners and get covered in the stuff, and when they wash it off, apparently that’s the one place it’s hardest to clean.’

‘So they actually work down in the pits?’

‘Yep. At the coalface.’ He chuckles at his pun.

So, wow. Don’t I feel stupid. This is, however, very exciting. Real coal miners, and I’m surrounded by them.

Not once in my entire time out here have I ever sat down with anyone I don’t know, while sober, and just started chatting. It takes me ten minutes to develop some questions and build up enough confidence.

After five separate chats, I’ve met five lovely straight men, and their five reasons for being coal miners are no different to most FIFO’s reasons for being out here. Money for saving, spending, family, debts or property.

None of them think their job is special in the slightest. Every question about it comes back with a single-sentence, word or grunt for an answer. When I ask about what they’re using their money for, however, they talk so extensively and excitedly about their non-FIFO lives that I need to interrupt them just to get a word in.

Day 18 – Saturday, January 26

Australia Day celebrations in town are cancelled because everything is soaking wet or underwater, and even the local horse races are called off, as the course is flooded. It’s the biggest annual event in town, and I only know that it exists at all after I overhear Dale talking about it.

‘Those bloody race tips I paid for from that professional punter bloke are now bloody useless. Fuck it.’

For everyone else onsite, it’s just another day.

Dispatch No 18 – Sunday, January 27 to Monday, January 28

Why would I know how to spell my Mum's name?

Day 19 – Sunday, January 27

One of the worst days I've ever had, and not just because of what I learn about the writing life.

I read half of the example user manuals Jonno has sent through, and realise that not one part of them will be useful. Also, I'm angry. They are so boring, tedious and pointless, it physically hurts to read each word.

These are documents that nobody will ever use, assembled from other documents nobody ever uses, probably only ever read by me and maybe a dozen others, although they've likely been printed hundreds of thousands of times.

Worst of all, they're all written in a way that's of no use to anyone, and this is the type of writing that pays. Well, it's the only type of writing that I'm being paid to write, and I want to be a writer.

Here are three random examples from other manuals. Now these sentences haven't been picked because they're especially and excruciatingly tedious, poorly worded or pointless. It's actually an accurate example of the type of language used on every single page.

Parabolic drawdown at large flows: *At increasingly large flows, a progressively increasing frictional resistance in screen setting and the aquifer will give a parabolic drawdown curve of the second degree.*

Quality Assurance: *Again the Quality Assurance plan is, at present, not site specific. Upon award of a contract the plan and associated ITPs are compiled to represent the documentation and specification requirements of the project in question. Once the necessary changes have been made, the plan in total must be submitted to the Managing Contractor for approval.*

Equalization Tank (T-1): *The equalization tank (T-1) is located inside of the WWTP at the head of the WWTP plant. The equalization tank (T-1) equalizes flow from the DNAPL recovery system and decontamination sump extraction pump prior to discharge to the OWS. Exact equipment locations and piping* *connections are shown on Drawings C-16A, C-16B, C-16C, C-16D, and C-17. Information regarding the equalization tank (T-1) is provided in the manufacturer's cut sheets in Appendix A.6.*

Now imagine reading hundreds of pages of this crap, then having to write your own, and try to resist digging your eyes out of your head with a plastic spoon, steel ruler or screwdriver – that's what on the desk next to me; you choose whatever's handy. Seriously, it is as if this stuff has been written by whoever does the furniture assembly manuals, with help from a group of dyslexics and the person who creates the clues for cryptic crosswords.

Late afternoon it gets worse, when Pando walks into the office.

'Xavier, what the fuck are you doing working in admin? You're a decent lump of a fella, and you look bloody stupid sitting there hunched over that tiny laptop. You're made for hard labour, outside with the real men. Not hiding in here like a little girl,' he tells me.

'Nothing wrong with paperwork,' says Donk.

'Mate from what I hear, you're man enough to do whatever the fuck you like,' Pando replies.

'What do you want?' I ask.

Pando throws a grotty folder full of papers onto my desk.

'What's this?' I ask.

'You're a writer aren't you? So write.'

I want to tell him to piss off, but Pando looks very much like a panda – big with soft edges and a placid expression, but I've seen video clips of when pandas attack and know Pando is prone to do the same.

'What exactly would you like me to write?'

'I don't know, I'm not a writer.' He stares at me. 'Come on, I dig the holes and lay pipe. You do the papers. Unless you want to dig the holes?'

From what Jonno and Dale have said he's apparently very good at it, and I really do want him to go away, so I open the folder.

'These are testing reports, but it looks like only about half of them have been filled out. So has the testing been completed? Like do you have all the results?'

'Xavier mate, I'm just a dumb tradey. Didn't even finish year ten,' Pando replies.

The only time these men are disparaging about their abilities is when they're trying to get out of doing something.

'But you still manage to receive emails about your flight details, and email head office whenever you think there's a problem with your wages.'

He says, 'Yes, but that's different because those are all things I want to do.'

'So have you got the results written down anywhere?' I ask again.

'Fuck this,' says Pando, and leaves.

An hour later, Jonno comes in.

'Did you help Pando fill out those testing reports?' he asks.

'I tried, but he wouldn't tell me the results.'

'Stupid illiterate prick. I'll make him come back tomorrow. That stuff needs to be sent off to Nuscon as soon as.'

Day 20 – Monday, January 28

This morning, Pando is waiting at my desk.

'What time do you call this?' he asks.

'I was at the pre-start meeting. Where were you?'

'I've got important shit to do. Now do these forms.'

He pushes them at me. On each page are a series of boxes, and I have no idea what's supposed to go into them.

'Also, it fucking reeks in here,' he says.

When the door is open, people complain about the heat, because the small air-conditioners are no match for the 40°C heat. When it's closed, they complain about the smell.

'Have you done all the testing?' I ask.

'Of course.''Did you write down the results?'

'That's why I'm in here. For you to do it,' he replies.

'So what should I write?'

'Fucking smartarse stupid cunt,' he says, and leaves.

Jonno comes straight in. 'Did you get it done?'

'No. I don't think he can remember the results.'

'Fuck! I'll send him back in – it has to get done today. Oh, and I need all the timesheets and JHAs for the last two weeks by the end of the day, and they need to be perfect. We're being audited.'

After gathering together the required daily timesheets, I flick through and forge any missing signatures. The guys know they're supposed to sign it every morning to verify their hours, and every day a few of them walk off without doing it. For the last few months, instead of wasting hours chasing them and since their hours are always identical, I've just added in the missing signatures myself. Only twice has anyone complained, and both of those times the guys had actually done the signing themselves, but had been too lazy to even remotely repeat what they'd always done.

The JHAs pose a much bigger problem, and not just because there are so many more of them. A JHA (Job Hazard Analysis) form is supposed to be filled out by each worker when they begin work, after a break, or if they switch tasks. While nobody sticks to all the rules, most of the guys do fill out a few each day.

The majority of those who work in mining and outside of an office perform highly repetitive and dangerous tasks for ten to twelve hours each day, often in extremely uncomfortable conditions, and as Dale explained to me, 'Most of the guys arrive at work and switch off their brains, if they were ever on at all, and they've got any brain cells left. So getting them to have a think about what might go wrong before they get to it isn't the worst idea.'

Part of my job is collecting the JHAs and making sure they're legible, before they're sent off to wherever it is they end up. Now the guys I work with can speak English just fine, but have some big issues with the written version.

This is an industry-wide problem for mining, and among the many examples I've personally seen and experienced of borderline unintelligible written communications, I also know of mining company project managers, with annual salaries of a quarter of a million, who've been sent to emergency remedial English courses designed for high-school students.

I try to keep up to date with JRT's JHAs because it can quickly build into a behemoth of a task and, although I last went over them three days ago, there are still already over a hundred I need to check.

Common mistakes the guys make include changing words that begin with 'en', 'im', 'un' or 'an' into 'in'. Examples include: 'invironment', 'inportant', 'inknown', 'incient', 'inpact', 'inage', 'inalyse' and 'inal' instead of 'anal'. Okay maybe not that last one, but if it did happen, it'd make my year.

They also have a lot of trouble with the word 'hole'. As in, a hole in the ground. Many times I've seen it spelled beginning with a 'w', lacking the 'e' and the 'h', and with two 'e's and all sorts of rogue silent letters. A few of the guys have so much trouble that they've given up on spelling it out, and now just draw a circle.

Then there's my name. A lot of the guys still call me Matty, as the name

'Matthew' was on the first high-vis shirts I was given and wore for three months until the ones bearing my name finally arrived, and they spell Matty with an 'ie', which I suspect is on purpose.

Damo even once told me, 'Of course it's Mattie with an 'i-e'. You know, like Pattie, because you've got a girl's dick.'

The best mistake I've come across wasn't on a JHA, but the application form of a guy who got a job with JRT. Under 'emergency contact' he'd misspelled his mother's first name, and it wasn't some fancy bogan attempt at being different. It was clearly wrong, as you should never spell 'Charlene' with the letters y, m, t or u.

What he actually wrote down was 'Sharleamtne'. Which looks like he just tried to include as many letters as he could, in the hope that some of them would be right.

So I had to call him up and ask what word he was attempting to spell, then how to spell it, because if you're calling someone's emergency contact that means something horrible's happened, and you don't want to notify the wrong person.

He told me, 'The name's Charlene, but I've got no idea how to spell it.'

'You don't know how to spell your Mum's first name?'

'Why the fuck would I? For my whole entire life, I've always just called her Mum.'

I've heard from other people in the industry that apparently Nuscon, along with other project management companies and the mining companies, often goes completely overboard with their use of JHAs. At some mining sites, Nuscon employees have to fill one out each time they want to use the photocopier or drive a work car home, and one girl at a mining company had to fill one out for picking up papers around her desk and putting up Xmas decorations.

One of the most ridiculous things I've heard happen, as a result of a JHA, involves a Debitel engineer who broke a shoelace during a site visit at a Western Australian iron ore mine. While she was in the storeroom picking up a replacement, the safety officer made her fill out a JHA. Under the column for 'action or procedure' you're meant to list how this hazard could be avoided in the future, and she wrote, 'By carrying a spare shoelace'. So from the very next day, the safety officer made it mandatory for the 800 people at that mining site to carry around a spare shoelace.

At a South Australian gold mine, the working-at-heights regulation kept being lowered by another overzealous official after a series of JHAs reported near misses because people mistakenly thought they needed to fill in all of the boxes, and it was okay to report what could happen instead of what actually did happen. The definition of working at heights was made so low that a delivery driver, who had a special reinforced and raised ute, was operating at heights every time he needed to get anything off the back of it. So every time he made a delivery to this particular mine, he had to be put in a harness and attached to a winch, which required a winch operator and spotter. The rope for the winch was two metres long, however, and his ute was nowhere near that high. Meaning the whole thing was pointless, because if he fell he'd still hit the ground.

Around 1 pm I finish with the JHAs, then just after 2 pm I hear Pando out in the smoking area swearing about what a prick I am, so I go outside and call him into the office.

'Fuck it's hot in here,' he says. 'Why leave the door open? And with the air-con still going. I thought you were supposed to be a greenie or some shit?'

I knew he'd say that.

I have the reports that have been completed spread across the desk. 'See these ones that you have filled out?'

'Yes,' Pando replies, while transfixed on the porn playing across the office on Donk's computer.

I shove a page in front of his face. 'See this? What would happen if I copied this onto the blank reports?'

'They wouldn't be right.'

'But would anyone notice?' I ask.

'Probably not.'

'Great, you're done,' I tell him.

'So I can go?'

'Yes.'

It takes about forty minutes to fill in the empty boxes and descriptions with my best guess at what the actual test results might've been.

Shortly afterwards, Jonno reappears.

'Here are the timesheets, the JHAs and those testing reports,' I tell him.

He quickly flicks through the three fat folders, then opens his mouth to ask me something. My guess is he's wondering how much I had to fudge to get it all done, to have an idea of the sort of trouble we could be in if caught, but instead he just closes his mouth.

Before leaving the office he says, 'You're a champion. Make a start on those manuals?'

'Not yet,' I reply.

I waste away the rest of the day skim reading the other half of the examples Jonno has sent. Again, nothing useful. At the same time I'm continually checking my email, going for pointless walks, reading online articles and otherwise looking for any distraction, and not just from the manuals.

A typical working life is full of tasks just like this, especially a FIFO life. People earn a weekly wage doing this type of thing precisely because nobody would do it unless they were being paid, and someone with money has decided that it needs to be done.

Nothing about what I want to do with my life needs to be done. It's me attempting art through comedy, which is widely regarded as one of the lowest artforms.

Which is another reason today has been so difficult – I've again been reminded that nobody needs what I'm offering creatively, so making a living out of it is always going to be that much harder, but I don't want to do anything else.

Fuck those manuals for making me feel like this.

Dispatch No 19 – Tuesday, January 29

The gluten-free dreamcatcher

Out here, you either become addicted to fitness or fatness.

Many who've previously shown no inclination either way expand and soften, or contract and harden, the longer they're trapped in the FIFO life. When I first arrived, I gravitated towards fatness. Unable to resist all-you-can-eat bacon every morning, cold cuts at lunchtime, and steaks and cakes every evening, I piled on the pud, at an average of 5 kilos a fortnight.

The days onsite are long, but for many they're not physically hard, just monotonous, and, as with most jobs that involve lots of sitting and not much thinking, people tend to overeat because they're bored. Which is especially easy when the options include ice cream and donuts. Then in the evenings there's little to do apart from eating, drinking and smoking, or going to the gym.

It's as if one of those putrid reality television gameshows has made it out here already, but instead of *The Biggest Loser* it's *The Biggest Winner*, and the race is to see who can have the first heart attack.

We all pack our own lunches and another common problem is that everyone takes too much, because it's free and you don't know how hungry you're going to get. Meaning the wastage is also out of control.

Last year, six weeks into my FIFO life, I'd even started to disgust myself. Surrounded by similarly expanding waistlines I felt like I was trapped in a land of *Teletubbies*, and was worried that I too would soon lose sight of my feet and penis. So I switched my allegiance to team fitness, which appears to be split into two camps – those that hit the gym because it's good for them and there are no other exercise options out here, and the few who lift weights because they're incredibly turned on by prominent veins, so are trying their hardest to resemble a muscle bursting with them. Another problem of being stuck somewhere so remote and working swings is that involving yourself in any team sport is near impossible and, stranded at the mining camp, you don't even have easy access to the town's local pool or anywhere worth running.

Mining-camp food is also renowned for its lack of variety, with guys often escaping into town for a petrol-station pie, a pub steak or Chinese food.

One night, a chef explained to me that the food on offer wasn't solely to blame.

'We keep getting complaints that the soup's grey, runny and tastes like shit. Also, that it's always the same, but there are two soups on every night, and always at least one new one.

So these complaints, they confused the hell out of me.

So I found one of the whingers and made him show me what he was on about, and I shit you not, he said soup, but he pointed at the gravy.

The bloody gravy!

He'd been eating bowls of gravy and thinking it was soup. What a unit.

Then every night, and I watch 'em, every night the guys drown their food in the bloody gravy, and then complain that everything tastes the same. Like gravy.

And they say there's not enough variety, but they always take a bit of each dish and it doesn't work like that.

You're supposed to eat like you would at home, you know like a meat or a fish and a couple of other things, then leave the rest for another night.

I'm telling ya mate, and you're one of the only ones who listens and I bloody thank you for that, but the fucking food here is only half the problem.'

Recent news stories have detailed the issues caused by all the unhealthy options on offer at mining camps and the rising rates of drug abuse, which includes steroid use. One study found that 83 per cent of FIFO workers at a particular camp were overweight or obese, more than 80 per cent had an increased risk of developing diabetes and heart disease, and that nearly 60 per cent of the meals offered included highly salty and fatty foods. Drug dealers have been found at other mining camps, some employees have spoken to the press about their addictions, and out here one night at the local pub I even spoke to a couple of government mental health workers who told me all about the parade of miners they see addicted to steroids and crystal meth.

Not many seem to manage tiptoeing between the freaks and the fatties, but I'm having a crack and this evening I was called out on it. At dinner they had tofu for the first time, and I got a bit over-excited so piled my plate high with it. Then there were no spare tables, so I sat with three diesel mechanics.

After chatting for a bit about whatever, one remembered seeing me perform at the Woodford Folk Festival. They then took turns bouncing insults off me.

'So he's a comedian? Is he funny?'

'I'll tell you what's funny, all that fucking tofu.'

'You're not some fucking vego or some shit are you mate?'

'What about gluten? Do you eat gluten?'

'I don't believe in gluten. Like I reckon it was made up by the fitness freaks.'

'Yeah, to sell us shit.'

'And make us feel bad for eating food that actually tastes good.'

'I thought comedians liked to talk.'

'And be funny.'

'Tell as a joke, will ya?'

'Sorry,' I told them. 'I'm not working unless you pay me.'

'No fun unless you pay. Just like a whore.'

'I bet you above his bed, he's even got a dreamcatcher that's gluten-free,' said the guy who was at Woodford.

'What's a dreamcatcher?'

'Hippie shit.'

'Why were you at Woodford? Are you a hippie?'

'Nah mate. Great place to take acid, which doesn't show up on any of the tests.'

'Is that true?'

'Hundred percent. I'm still working here, aren't I?'

'So am I, and I take plenty of stuff other than acid.'

Once I finished my food I told them, 'Great to meet you guys. Thanks for the seat.'

Since I know from experience that they probably thought they were being hilarious, and had no idea that they were actually cunts.

Dispatch No 20 – Wednesday, January 30 to Saturday, February 2

You're not drinking? You're a dickhead

Day 22 – Wednesday, January 30

So I'm not going to have one single drink in February.

Febfast is coming up. Which involves abstaining from a range of possible things, but alcohol is the main one, so that's what I'm going with. Verity suggested it. However, I'm not doing it for her.

More accurately, I'm not just doing it for her. I've got plenty of other reasons.

Anybody who knows me will be reading this through a smirk and thinking, 'Bullshit'. Which is fair enough. For years I've been the last man standing, the good-times guy, the person always ready to turn a few drinks into a full-on bender.

While I've only been drunk once in the last two weeks, in the last three months I can count all my alcohol-free days on one finger, and I don't even need that finger. Every night I have a beer after work, and it's always delicious.

The recommended daily maximum alcohol intake for an adult is two standard drinks, and even on my quietest night I still exceed that, so while I don't drink much compared to most out here at the mining site, by any medical definition I'd still be considered an alcoholic.

Which is a real worry.

I'm already well aware of all the health problems caused by alcohol, but there's one massive point missed by all the campaigns against it.

Drinking is fun.

People don't drink for the side effects. For the weight gain, hangovers, to destroy their livers, hearts and brains. People drink to make good times even better. Or like me at the mining site, to make terrible times bearable. Alcohol is actually so accomplished at doing this that many people know the side effects and continue to drink too much anyway.

I'm trying to get back to being a full-time artist, while working out here to erase my debts. To do both with any semblance of success something has to go, and that something is drinking.

Out at the coal mine it won't be that hard – alcohol is an integral part of socialising but I still don't have any friends out here. Thus my problem of not having any friends is no longer a problem. Many of the men out here even act as an inspiration – all their drinking and smoking and coughing and wheezing and sweating while sitting inspires me to be nothing like them.

What will be difficult is getting through the Adelaide and Perth Fringe Festivals, as both happen in February and I'll be onstage attempting to be funny two or three times a day.

Over my entire comedy career I've easily done over a thousand shows, and guess how many of them I've done sober? Without a drink in my hand, and one before and after the show? Zero.

Without alcohol, I would've never gotten on stage to begin with. For my first time, I downed jug after jug, then hit the stage at that magical point where you're a flawless pool player, and you know the perfect thing to say to the girl. For my second, I employed a similar strategy but was on towards the end, and when I finally made it onstage I saw three microphones, not one. It didn't go well.

Also, when not drinking I still have the urge to write, but I doubt I'd have much to write about.

'Today I went to work, watched a movie, and went for a run.' Bor-ing.

'Last night I got drunk, threw eggs at a nightclub queue, which all missed, traded five cigarettes with a homeless person for a tab of acid, and woke up in a park next to a girl and a dog.' That's better.

Almost everything's better when I'm drunk. Music festivals, friends' birthdays, family gatherings, nightclubs and first dates. Even the mining site – but even better than that is being sober and not here.

Alcohol is my warm security blanket of self-confidence that insulates me from the doubt and second guessing. Especially important for a single, male comedian, as I need it when onstage to talk to the audience, and then afterwards to talk to the girls. This time around with Verity I've already got the girl, and probably wouldn't be able to even attempt this without her.

So I want to break my dependence on alcohol, to extend my life for as long as possible, and to achieve more in the time I do have. No more wasting days hungover and hating myself. For a month, anyway.

Also, I love writing and performing. When drunk, hungover or even slightly impaired, my writing is rubbish. Then the performing – while that works fine with alcohol, I've never given the alternative a go. So here's your big chance, non-drinking Xavier.

Really, I've been thinking about doing something like this for a long while.

So as you see, I'm doing this for me. As well as Verity.

Day 23 – Thursday, January 31

Tomorrow there's an RDO, and for the first time in my mining life on the night before an RDO, I don't go out drinking. I know Febfast hasn't started yet, but I just don't want to. Which makes Verity very proud.

Before bed last night and this evening, I wandered around the campsite and noticed how many people had drinks in their hands. Alcohol is such a big part of mining and comedy that anyone who doesn't drink stands out far more than those who do.

Which frightens the freak out of me, because I detest being the centre of attention unless I've got something to say, I've rehearsed saying it, and I think it's worth sharing. I can't stand undeserved attention, and that's one of the reasons I detest my own birthday.

When I'm onstage people have paid to listen, and it's an opportunity that I've worked hard for so I feel like I deserve it. With birthdays, all I did was get born. Really, they should be all about mums, as they did all the work.

So to avoid the attention that comes with refusing a drink at the pub, club, house warming, barbecue, wedding, engagement, birthday party, lounge room, office or comedy club, I've always had one. It's just easier.

Among most people I know, when someone isn't drinking we suspect that they're pregnant, and we're usually right. Then the guys I know are never not drinking.

Also, alcohol makes me an extrovert, which people like.

Sometimes when I arrive at a party sober, or if I'm driving, people will ask me what's wrong, or why I'm being so quiet.

'Nothing's wrong, I just don't have anything to say,' I tell them.

'That's not like you.'

I'll quickly down some spirits and forget about driving, just so I can go from normal Xavier to fun Xavier, and it always feels good to be part of the party instead of just an awkward bystander.

Maybe that's a bit sad, but it's the reality, and a big part of the reason I want to stop drinking is to see if I can find that fun Xavier without the booze.

So I don't want to bang on about it too much, but this is going to be really tough. I actually don't think I can do it, which is another reason I want to try.

Day 24 – Friday, February 1

In the early afternoon, after spending hours on comedy admin and even writing a few jokes, I borrow Ben's ute and drive into town to visit the Workers Club. Just to see what it's like to sit in a pub and not drink.

It sucks.

What doesn't help is that some JRT guys are there.

Within two minutes, Cliff asks if I'd like a beer. Which

is lovely. In different circumstances, we might've even been friends. I follow him over to the small table where he's drinking outside with Dale and Pando. The table's covered in empty glasses, losing horse-racing tickets, and two full ashtrays.

'What about Xavier? Does he want a beer?' asks Dale.

'Not anymore,' says Cliff.

'What's wrong with him?'

'What? Apart from the obvious?' asks Pando.

'He's doing that not drinking for a month thing. Dry Feb or something?' says Cliff.

'No friends Feb,' says Dale.

'Being a dickhead,' says Pando.

'I've actually done it too, but back when I did it, it had a different name. What did the judge call it? Oh that's right. Probation,' says Dale.

'Mate, if you want to stop drinking, why make such a big thing out of it? If I wanted to stop, I'd just stop. Just like that,' says Pando. Who then scratches the underside of his spherical stomach, downs the last half of his beer, and lights up another smoke.

He's got a point though. If I want to stop, why the big parade?

Well I've tried doing it quietly, more than a few times. During the week I'm fine, which lasts all weekend if I don't see any friends or family, and don't go outside or leave my room. As soon as I do though, and someone offers me a drink, I never refuse.

By making a commitment to others, saying it out loud, signing up, paying a fee and maybe raising some money, hopefully that'll work. What's really working though, is Verity. If I don't manage to do this I know she'll be shattered, and it might be enough to end things.

Into the carpark I find reverse without a problem, return to the campsite and churn out a few more hours of comedy duties. It's the first RDO where I've achieved over half the tasks I'd planned to get done.

Day 25 – Saturday, February 2

Excuse me for a moment, while I go and make my eighth coffee for the day. All before lunch too, and I'm not even on track to beat yesterday's record.

Anyway, this whole not drinking thing has been easy so far, but it has only been two days, and it's not as if I've just replaced alcohol with some other vice. Also, since I stopped drinking, I've started shaking and have hardly slept, but I think that's more to do with all the coffee.

There haven't been any actual alcohol cravings, and really, it's been no trouble whatsoever because I've simply avoided it. Apart from that quick cameo at the Workers Club last Friday, where I certainly wasn't sitting there telling myself, 'Look at those poor fools ruining their livers,' while secretly thinking, 'I would tongue kiss any of them, just for a taste.'

I think I need another coffee.

Basically, I've been pretending that alcohol doesn't exist. Then when that doesn't work, drinking the odd ginger beer, because it says 'beer' on the label, and I'm trying to trick my subconscious into believing that it's the same thing.

Which hasn't worked either. I don't even like ginger beer.

At the campsite that night, Jonno discovers that I've stopped drinking.

'You don't want one? You're on the wagon? Good on you. Why you'd want to do that though, I've got absolutely no idea.'

Donk adds, 'I cut down on my drinking, so I could afford more smokes.'

Jonno cracks another one open, and it makes that crisp, clear, slightly moist sound particular to a perfectly chilled beer.

'Why would you want to give up this? Best part of the day,' he says.

I try to reply, but am slobbering too much to make words.

The after-work beer is one of my favourites. Up there with a beer with my Dad, the slightly illegal driving to a party beer, afternoon watching the footy beer, the Sunday morning hangover-curing beer, Sunday morning still-out beer, the back at a girl's place for the first time beer, and catching up with a mate beer.

After eating dinner alone, as per usual, I walk directly from the dining room to my room, and over the trip count thirty-seven people smiling, laughing and drinking.

I feel like I'm in black and white, while everyone else is in full colour.

Dispatch No 21 –Sunday, February 3

A horse walks into an office

For Dale every Saturday, horse racing takes precedence over work. Yesterday was no different, and just after morning tea, he sat down in the office and continued scanning the form guide.

'Aren't you supposed to be on a deadline or something?' I asked.

'I've been so busy I haven't even had a chance to fart, let alone read the paper or do the fucking form.'

'I notice that you never do the form when Jonno's around,' I replied.

He smiled, and without looking up said, 'Keep talking like that and you'll get a slap.'

Dale used to share his race tips around, but after a month of only a few winners, anytime someone mentioned the poor results he'd say things like, 'Want a smack in the mouth?' or 'If you're going to complain, I'm not going to fucking give them to you'. Then that's exactly what he did, and nobody's had a tip since.

Yesterday, however, Dale had a win in the thousands, so Sunday's become all about the races as well.

During the call of race three at Moonee Valley in Melbourne, Roy from Nuscon comes in and asks me about some safety documentation I've already sent. I email it to him again while he's standing there staring at the radio – which is something else that can get you instantly sacked.

Once the race is over he says to Dale, 'Have a win?'

'Rarely.'

'So you follow the fucking nags pretty closely then do you?'

'You could say I've got slightly more than a passing interest,' Dale replies.

Roy then starts speaking in a hushed and hurried way I've never heard him use before, barely even swearing as he takes over an hour to tell Dale about the horses his brother helps to train, and the day they won over a hundred thousand dollars at the track. Next, he gets me to play a DVD that he just happens to have on him, which contains the first two races of a particular three-year-old.

'Thing is,' he tells Dale. 'We've got a couple of shares in this filly still available, and now's the time because she's just about to start pulling in some very impressive prize money.'

'How much are the shares?' Dale asks.

'Only $30,000 for one-eighth, and there are only two left.'

'That's pretty reasonable.'

Roy has been slowly moving toward Dale, and is now so close they're basically touching.

'So what do you reckon?' Roy asks.

'I'll have to speak to the boss,' says Dale, and all those in the office, Roy, Donk and myself, know that he means his wife, and not Jonno.

Roy still appears positive, although even I know that saying you have to check with your wife first is man-code for not being interested.

After ten more minutes of enthusiastic chatter Roy leaves, and Dale asks Donk and I what we think of Roy's proposal.

'If he's won a hundred grand in a day and it's all going so well, then why the hell does he need a job out here,' I say.

'The only guys I've seen more eager than he was are the blokes in line at the brothel, waiting for their first one after three weeks onsite,' says Donk.

Dale shakes his head. 'Yeah, he was pretty keen.'

'So are you interested?' I ask.

'Fuck no.'

Less than two hours later, Roy's back.

'Only one share left!' he says. 'If you want it, I'm going to have to get a deposit off you today. Otherwise I won't be able to hold the share for you.'

'How much is that? The deposit?' asks Dale.

'Three grand.'

Dale whistles. 'Sorry mate, I think I'll have to let this one go.'

'Okay,' Roy replies. 'I can probably hold it for a week.'

'Wait as long as you like, but my missus is never going to go for it,' Dale tells him.

Roy's shoulders, gaze and mood all drop to the floor, and he leaves.

'Out here there's a lot of blokes addicted to shit, and he's got it as bad as anyone,' says Dale. 'The way he was trying to reel me in, my guess is that he needs the money because he owes someone.'

Dale's very likely right. At the pubs in town every evening FIFOs and DIDOs can be seen pumping money through poker machines, over the bar, and into horses. I've also spoken to plenty in mining who had grand plans of saving big, buying a property and setting themselves up, but after years in the industry haven't put away a single cent. Instead, they've wasted it away on a combination of gambling, drugs, alcohol, cars and girls. Just as men have done all through human history.

Dispatch No 22 – Monday, February 4 to Sunday, February 10

One never-ending week

This is every week on the mining site, compared to a standard city job.

Day 27 – Monday, February 4

On a Monday in the real world, I'm recharged following a weekend away from work.

On the mining site, it might be Monday but you may as well just call it 'day' because they're all identical.

The only day that matters is the day you fly home for your break. Ask anyone, and they will instantly be able to tell you when that day is.

One difference is that head office is back on the phone, interrupting my progress on comedy admin and other career- related tasks.

Three hours is then spent chasing up everyone who didn't fill in the timesheet or the daily safety report properly at any time over the past week.

There's also always a rumour circulating that someone or everyone is getting fired, one company out here is about to fall over, the job is about to finish, or we're all getting a free backpack or barbecue.

I text Verity around seven times, and she texts me twice.

Donk says, 'This time, I've finally found the local sex club! It's around the back of the police station.'

He's never found the local sex club.

Day 28 – Tuesday, February 5

At my city job, I'm still spry from the weekend. Since I enjoy what I do, I'm also looking forward to spending the next four days knocking things off my to-do list.

On the mining site, I'm unable to remember what day it actually is. Even after checking twice.

There's the daily breath test, the site-wide meeting, stretching and the company meeting, where I hear or smell at least ten different men fart.

Jonno congratulates us for doing something very well.

Every week I also try at least one energy-saving measure.

‘I think we could turn down the air-conditioning when it’s not that hot outside,’ I say.

‘No,’ Donk instantly replies.

‘Stop thinking,’ says Pando. ‘Nobody cares what you think.’

One of the printers is out of ink. One of the photocopiers is jammed.

Every day, at least three people ask me about their flight in or out. At least one of these questions will turn into an argument, during which I say I can’t do anything to help, and after five to ten minutes of yelling and swearing, I still can’t do anything to help.

A delivery guy arrives with some supplies. He’s at the wrong worksite.

A JRT employee commits an offence that could get him fired.

This week it’s Cliff, who was caught drag racing some electricians on the way back to camp, while driving a JRT work ute, and carrying four other JRT employees. He very nearly slammed into a semi-trailer coming the other way, and that’s who reported the incident.

Dale tells me all about it, then laughs and says, ‘He’s just bloody unstable, that Cliff.’

The options for Dale and Jonno are to either issue a verbal warning or a written warning, dismiss or transfer the employee, or do nothing.

Unless forced to do otherwise by Debitel or Nuscon, they usually do nothing.

I call Verity. No answer.

I call my parents, as they almost always answer.

I fall off the treadmill.

Day 29 – Wednesday, February 6

It’s the mid-week hump at my normal job, which I get over by meeting a friend after work for a run then dinner.

Onsite most get over hump day by drinking and smoking after work. Same as they do every day. Apart from those who are flying out, and ecstatic to leave. Or those flying in, who are the opposite of ecstatic. If asked about their week away, those returning all repeat at least two of these six statements:

1) The whole week went so fast, it felt like one day. While out here, every day feels like a week.

2) I took so many drugs that I only slept once. So it was exactly like one day. It was awesome.

3) I was drunk the entire time. It was awesome.

4) My kids are doing so well. It was awesome.

5) I did a ton of work on the house. It was awesome.

6) I did very little. It was awesome.

Dale whinges about his wages, because although he's the onsite supervisor, he's paid only a tiny bit more than everyone else.

The JRT employees get me to do non-work-related tasks including: faxing their divorce lawyer, confirming an already confirmed personal flight, changing a password for an online gambling service and proofreading a resumé, among many others.

Two standout requests this week were:

Donk asking me to organise a flower delivery. 'My wife usually does all the internet stuff.'

'She'll love them,' I tell him.

'They're not for her,' he replies.

Jokka, who claims to have hurt his back, so requires an urgent appointment at the chiropractor. I give him a piece of paper and tell him, 'Here are the numbers for both places in town.'

'Can you do it? I haven't got my phone.'

'You're holding it in your hand.'

'But we're not allowed to use them onsite,' he replies.

I check through Jerome's email account and folder of deleted items for anything important and find plenty that he's missed, ignored or erased.

While making a coffee, someone tells me there's a coffee machine in the Debitel office.

Verity texts me three times, I text her once. I call and she answers. We chat for ten minutes.

Day 30 – Thursday, February 7

At my nine to five job I spend the morning considering options for the weekend. Also, I realise that I'm not going to get anywhere near as much done this week as I'd hoped. So I procrastinate by updating my to-do list with new tasks, and ticking off what I have achieved. In the evening, I go to see a band, a movie or, if I'm lucky, on a date.

Onsite there are so many different ways that I experience time. It can be slow or hurried. Some days it stretches, collapses and leaps around, all before morning tea. More than anything, it's a constant and oppressive presence, because although it never stops, it also feels like it's never going to arrive anywhere as every day feels just like the last and the next.

Jonno asks me to go outside and sketch, measure or check something. I get lost looking for it. If I run into anyone, they always ask me what I'm doing, and I used to think this was because they suspected that I didn't do anything. While I still think this, I've also realised that it's because they're worried I'm checking up on them, and might figure out that they're not doing much of anything either.

An ex-employee or their partner phones about outstanding wages, or a current employee slowly runs me through their pay slips as they're unsure about something. Only twice have I ever been able to help.

Someone has had enough and wants to quit. They never do, but might turn up drunk in order to tentatively explore the option.

I fall off the treadmill. Again.

I text Verity once.

I stare at my phone, and scroll through my contacts, trying to think of someone who isn't my parents and would answer a call from me.

Day 31 – Friday, February 8

Last day of the inner-city five-day working week. There's a cake for someone's birthday, I finalise my weekend plans, spend too much time online reading lists with hilarious headlines that are otherwise disappointing, and crack my first cold beer at a minute after five.

Out at the mine, I wake up and realise I'm still here, and breathe out heavily. Then decide to skip the gym and sleep for another forty-five minutes.

Jerome does something incredibly dumb.

This week, Debitel is stressing about some deadline, so have made it a rule that, for the time being, no onsite contractor is permitted to reduce their workforce.

Ned and Jokka, however, are being moved to a site that's even further behind, and Jonno made it very clear that we were to keep this very quiet.

Less than an hour later, Jerome has plastered both of our offices and the crib room with posters inviting everyone to a barbecue this Saturday night, 'For our dearly departing friends, Ned and Jokka!'

Dale and Jonno are so mad that I don't make any jokes about Jerome's unfortunate word choice.

A delivery guy phones me. He has some supplies for us, but he's at the wrong mining site.

There's a safety panic, which for a moment seems like an absolute disaster, but is solved once the paperwork is either found, forged or the whole thing is forgotten about.

Someone phones about an unpaid invoice and I direct them to head office. Rarely is it the first time that they're calling.

We need to depend on a local to do something, and are let down.

It's a water tank delivery this week, and at midday the driver says, 'It's Friday. I knock off at 2 pm on a Friday. So I won't be able to do the other loads until Monday.'

'We're here until 5 pm,' I reply.

'Sucks to be you.'

Verity texts me once. I text her back.

Later that night, I stroll past a group of JRT employees in a circle, drinking, smoking and chatting. They quieten as I approach, and once I've passed, they laugh loudly.

Day 32 – Saturday, February 9

In the city it's the weekend. Hooray! So I'm either recovering from a hangover, or up early to do something amazing, because I'm in the city and there are options galore.

On the mining site, the only way I know it's the weekend is that Dale has the Saturday horse racing blaring out of a radio.

Jonno berates us for doing something very poorly.

Someone misses the morning breathalyser. They explain to me during morning tea that they arrived early to finish off something urgent, and I tell them to wait a couple more hours before going over to Debitel to do it, as I can clearly smell alcohol on their breath.

An employee from JRT or some other contractor comes in selling $2 raffle tickets for his child's school, sporting club, or a charity.

Most buy a few, and nobody ever points out that just a fraction of the seller's weekly wage could buy all the raffle tickets several times over.

I always wonder if it's just me who thinks this.

Verity and I talk for over an hour. Thank goodness.

Day 33 – Sunday, February 10

It's still the weekend in the city, tinged with a little sadness because I'm back at work tomorrow. I stay out too late, because there's fun everywhere.

Out at the coal-mining site, it's as if the days are a liquid being poured into a vortex where they vanish, and then I'm older.

There's another argument about flights.

I text Verity twice, and hear nothing back.

Dispatch No 23 – Monday, February 11 to Wednesday, February 13

Girls, girls, girls

Day 34 – Monday, February 11

I call Verity, leave a message, and follow up with two texts.

I'm a little over-excited. After five weeks, it's only three sleeps until I meet her in Perth for Valentine's Day. She couldn't afford the flights and I didn't mind paying, since I need to be there for shows anyway. It'll be well worth it, and if things work out how I hope, eventually we'll end up sharing every cost anyway.

Around midnight I'm asleep, and she texts back with, 'Sorry. Work has been crazy as!'

Which is all cool. Lawyers do work very hard, and that's what she's working towards.

Day 35 – Tuesday, February 12

Verity phones at 4 pm, while I'm at work, and I go out into the heat to talk to her, as I don't want Donk, Ben and Dale listening in. They know very little about me and have never asked, so there's no point letting them in on anything.

It's over 40°C, but I've been waiting for this call since yesterday.

'Two more sleeps!' I tell her.

'Is that right?' she asks quietly.

'Are you okay?'

'This distance thing. I don't think I can do it anymore.'

Everything sinks. 'It...it's only one more day.'

'But what happens after that?' she asks.

'What if I quit? I've been thinking about it.' Not by choice, but there's no reason she needs to know that.

'No!'

'Why don't we see what happens in Perth, and take it from there? It's only two days away, and even if we decide to call it off, we both already know how well we get on together. So whatever happens, we'll have a good time.

Every which way you can imagine.'

I'm talking too fast, and that last sentence was supposed to be flirty, but instead I feel like I'm trying to sell her something.

'I just don't think it's a good idea,' she says, reverting back to her meek tone of earlier.

'Well I do.'

'Okay,' she breathes out heavily. 'I've met someone, and we're going out on a date tonight.'

'So what are you saying? That you want an open relationship? I suppose that could work, until we're serious. If things end up going that way.'

I couldn't hate the idea more, but I've got to get her to Perth. I'm sure we'll have such a wonderful time together that afterwards she won't hesitate when I again suggest being exclusive.

'It's our fourth date,' she replies.

'Oh.'

I notice that today in the paddock behind the mining site, I can see six cows. Two more than my previous record, from back before Xmas.

'I'm really sorry Xavier. You're a great guy. Like really, you're wonderful, but I can't do this. We hardly even know each other.'

When Verity doesn't immediately hang up, I think it's because she's still interested and only needs to be talked around. So for half an hour we talk in circles, my phone getting so drenched in ear sweat that she becomes difficult to hear.

As I try to convince her from every angle, I come to realise that she's probably only still on the line because she feels guilty. Of course it's over – talking is only one part of a relationship, and certainly not enough by itself to sustain a new one. Well, not this one anyway.

I let her go as it's nearly 5 pm and I don't want to miss the bus. Maybe I should've asked for the money for that return flight? It was like $600, but that's not me. Even if she offered, I wouldn't take it. If she insisted, however, well I might accept it then.

Back at the campsite, I sit in the bar with a glass of water, hoping someone will talk to me but lacking the confidence to start a conversation with a stranger. I call my five male friends back home. None of them answer, which is fair enough. We call each other maybe twice a year, and they've all got families now.

I feel so far away from everyone and everything. I haven't been touched by another person in nearly five weeks, or even had a decent conversation with anyone who's not Verity or my parents.

Out here, most of the men are already so much further along with their lives than I am, and some of those with young families are a decade younger than me. When they hear that I'm 34 and single, with no girlfriend, house or assets, they look at me like I've failed. Most of the time I suspect that they're right, and in this very moment, I'm convinced of it.

Maybe I should talk to a professional, but I never have before, and I'm not about to give in now. Also, it's not as if I'm depressed enough to harm myself or others. Professionals are only there for people with actual problems, and although this feels acute and special, I know it's just the regular kind of heartbreak.

I'd really like a drink and although I'm in a bar, I know I won't have one. I'm going to make it through this month without alcohol. It wouldn't help anyway – they only serve mid-strength at the campsite bar, and close in half an hour.

One of my aims this year was to analyse how FIFOs manage their relationships while onsite. Now I didn't expect it to be such a raw first-hand analysis, but there you have it. They suck.

Many of the guys I've spoken to have had similar trouble sustaining or starting relationships while FIFOs, and the men who aren't married often have at least one ex-wife. I even asked a few guys if they thought FIFO work was a factor in their divorce and was repeatedly told, 'FIFO is the reason my marriage ended. Is that enough cause for ya?'

According to others, shorter swings such as two weeks on and one week off almost always lead to a happier home life, especially compared to the three or four weeks on and one week off most people out here are working.

Anyway, now that I'm single and still suck out here, what are my options?

Well, I don't live in Adelaide or Perth, so anything that starts while I'm there performing will finish as soon as I leave.

There are about a dozen women currently at the job site where I'm based, and maybe fifty or so out of around a thousand back at the campsite; and I've got no idea how many are single or even interested in dating a FIFO. So really, it may as well be zero. Soon enough I plan to be out of this for good anyway, and I'm sure the women out here don't need another drooling man pouring on the charm because he's lonely, desperate and bored with moisturiser and tissues.

The few FIFO women I've spoken to said that despite all the men they never feel threatened, but that every time they move, they do feel eyes on them. One female environmental officer said to me, 'It's like being onsite with a whole lot of big brothers, who'd gladly give you one, if you asked.'

I've also been told that on a different mining site, all the women were advised to avoid eye contact and situations where they were alone with one or two men, and to talk about their partner as often as possible, even if they didn't have one.

There are the rare couples who met out here, or were lucky enough to get jobs based at the same campsite and, although there's no double-bed donga, they must sort something out because it definitely goes on. In my teenage years and early twenties when a single bed was the only option I always made it work, but I might struggle to convince a girl it's the done thing now I'm in my thirties. Some couples are unlucky enough to be on different swings, and are forced to spend their week off away from each other, which to me just seems unnecessarily cruel.

So hook-ups do happen out here, but a lot less than I overhear men talking up their chances. With some guys it seems that whenever they interact with a single girl they reckon they're definitely in, even though these men never seem to get anything apart from a kind word, a smile or an email reply.

Donk has then spent weeks searching for the websites set up so that FIFOs looking to cheat can get connected both online and off, but has never found one, which strongly suggests that they're only a rumour. It must happen, but I doubt it's organised enough to have a web address, and I wonder how many of the men I'm working with are currently being cheated on. Well, I know of at least one – me – but not anymore, now that I've been dumped.

I've heard multiple stories about people who have their onsite relationship and an offsite marriage; however, once again I'm very light-on for actual evidence. One bloke I knew had all the dating apps on his phone and regularly scoured the local area, 'you know, just to keep up with what's available', in between receiving photos of his newborn daughter from his partner.

In terms of sex that you can pay for, every rostered day off out here a procession of cars and mini-buses heads to the nearest city, around three hours away, for its range of strippers, massages with happy endings and brothels. Frequent justifications for this sort of thing include, 'there's nothing wrong with reading the menu as long as you eat at home', 'it's not cheating unless you stick it in', 'it's not cheating if you pay for it', 'it's not cheating if she never finds out', as well as, 'who gives a stuff? She gets her jewellery, I get my fucks'.

I was invited along but never went, and also knocked back the couple of offers I got to hang out with a group of guys to watch 'some seriously awesome and fucked-up porn'. For some reason the idea of sitting in close proximity to a whole bunch of men with erections didn't appeal. There's no doubting that, apart from all this, plenty of personal porn watching also goes on, and one evening I had a chat to one of the cleaners who reported that she'd come across, 'Plenty of rubber vaginas and fleshlights, dildos in both the guys' and girls' rooms, a collection of rubber nipples, and oh-so-many bins overflowing with used tissues.'

There are even men who tell their partners they need to work through their week off, then take off to Thailand for a week packed with drink, drugs and sex – and this is one I did hear about often from the guys that did it.

Despite all this, the large majority of guys don't seem to get up to anything untoward, or are clever enough to keep quiet about it.

So considering all that, it seems that I'm out of human options unless I'm prepared to pay for it. Which is something I've never done, and although I'm desperate for someone to touch me, I'd much prefer a hug to a rub and tug.

Day 36 – Wednesday, February 13

At work for half a day as I'm flying out this afternoon, and I should be more excited. I've been looking forward to this for five weeks, but the main thing I've been anticipating is now gone. I feel like I'm about to open what I thought was going to be a premium wine only to be told, just as I'm about to uncork it, that the bottle is empty and I'm unable to get a refund. So I suppose there's nothing left to do but open it, and pretend to enjoy the stale air.

If I was going to snap and have a drink, it would've happened last night. It didn't, and back at the start of February I said I wasn't doing this just for Verity. Now, I get to prove it.

As I'm about to leave the office and jump on the minibus back to the campsite to get changed, pick up my bag and head to the airport, Jonno comes in and says, 'Got a second?'

'Sure,' I reply and follow him out to the smoking area.

He lights up, and offers me one for the first time ever. When I refuse he says, 'That's right, every time you smoke, you spew.'

I force a laugh, and he doesn't even smile.

'So I'm going to need you to pack up all your stuff,' he continues. 'As you're away for two weeks instead of one, your room will go to someone else, and I don't know where you'll be staying when you come back. If you come back.'

'If...what? I'm sorry?'

'Things are slowing down and in two weeks, I don't know if we're going to need you.'

'Oh. Okay. No worries.'

I want to ask if there's a problem with my work, but I could name several, the most obvious being that I don't do very much of it. If I'd accepted that job offer this certainly wouldn't be happening.

'When will you know?' I ask.

'About what?'

'About whether or not you'll need me back out here.'

'I'll give you a call. Thanks by the way, for everything so far.'

Again he's not smiling, so I assume he's not being sarcastic. Which means I might be okay.

I really, really hope so. I'm still over $15,000 in debt.

Dispatch No 24 – Thursday, February 14 to Tuesday, February 26

Beer made from angel wings and unicorn tears

Perth Fringe World, Days 37 to 44 – Thursday, February 14 to Thursday, February 21

Tonight is my first comedy performance in over a month, and I'm single, sober and it's Valentine's Day. Couldn't have planned it more poorly if I'd tried. Nearly wish I was back at the mining site.

Sure, Valentine's Day is an overblown circus of consumerism, but it's everywhere. Also, it's a day that focuses on love and getting laid. Never a bad thing. Instead of once a year, however, maybe we should make it every Wednesday? One way to get over the mid-week hump. Literally.

While I'm no longer doing this month without alcohol for Verity, I'm now doing it in spite of her, which I'm finding to be an even more powerful motivator. It might even win her back – I'll call her at the end of February. You know, just to check in.

So every show I'm doing over this two-week period will be in a venue that serves alcohol, performed to people drinking it, and before and after every show I'll be hanging out right next to it. In bars. It's an occupational necessity.

I imagine it'll be like breaking up with your girlfriend, then moving into the spare room of the tiny apartment you two will continue to share, because nobody can afford to leave. Then every night, you sit alone on the couch watching repeats of M*A*S*H while she parades a new suitor through to your old bedroom and the king sized bed that's half yours. Where they give it an extremely audible workout. Now I'm not saying that any of my previous girlfriends were promiscuous, but alcohol, she's happy to get inside anyone.

So after ten years and over a thousand shows, my first ever performance completely sober goes okay.

I began with a new joke, which bombed, and another, and another, and I was dropping so many bombs that I felt like I was waging war, not doing stand-up comedy.

Then came a laugh, followed by a big laugh, and a proper big laugh, and a huge laugh. Despite the positive response, the voice inside my head second-guessing my every action was also particularly loud throughout, and that's something else I usually deaden with drink.

Afterwards, several audience members came up and told me how much they'd enjoyed the show. Some offered to buy me a drink.

Which I refused.

Reluctantly.

As it is well known that after a gig, drinks bought by pretty girls are the most delicious of them all.

The show certainly would've been better if I'd had a drink, or seven, and the smart thing to do would've been to build up slowly. A sober five minutes, followed by an abstaining ten, and on until I was ready for an hour. It's far less fun, however, to slowly climb a cliff, as opposed to just jumping off.

Then the next night, things get ridiculous. As I'm booked to perform at the Little Creatures Brewery in Fremantle, where all performers get unlimited free drinks.

The promoter tells me, 'Here is liquid gold. Made with hops grown in fields of clouds, yeast cultivated in the very bosom of Mother Nature, and ingredients including angel wings and unicorn tears. All cooked in a fire started by the final breath of the last dragon. Drink your fill, and be engulfed by its wonder here and now, for you may not consume any outside of this magical place.'

Or that's what I hear. When I'm stressed, I spend most of my time in a fantasy land that's part medieval England, part Narnia. Alternatively, I might be seeing things as they are, as Perth is a few years behind the rest of Australia.

By the way, if you're from Perth and reading this, well done! You know words!

Seriously though, I'm just doing my bit to keep it between you, me and the over two and a half million people who live in Western Australia how wonderful, progressive and reasonably priced the place genuinely is.

Instead of a beer, I have a four-shot coffee. Which is what I'm at now, just to feel even the slightest buzz, and I reckon I'm one latte away from snorting the stuff.

My decreased booze intake has also already resulted in a little weight loss. While being fat can help with the funny, I'm now single, and it's much harder getting a female interested if you're the size of a house, and sweating like an elephant-sized stick of butter in a sauna. Unless you're John Candy, John Belushi or Fat Albert. Who are all either fictional, or dead.

The show goes well, and a couple even approach me afterwards, and offer me a threesome.

I wish. Not because I'd be up for it, but it's always nice to be asked.

What actually happens is that the couple tell me, 'We're doing Febfast as well, and have been reading your blogs. We're struggling too but we'll make it, and think you will too.'

Which made me realise why big gestures like Febfast are so important.

I don't have a problem exercising or eating well most of the time. So when people go on about diet clubs, Zumba and that morning commando one where you pay some ex-army guy to scream at you until you spew, I see it as a weakness. If you're not mentally strong enough to do it on your own, how is wasting money to be part of a group going to help?

You want to lose weight? Eat less, exercise more, and exactly the same could be said to me. You want to stop drinking? So just shut up and stop.

However, alcohol isn't something I could give up on my own. It's just so much easier not to, especially when you're surrounded by it, everyone else is doing it, and you depend on it. Which is why it meant an incredible amount when that couple came up to me. I know my problems are pretty pedestrian, but I'm struggling and even just those few words really helped.

Alright, enough with the touchy feely shit.

Point is, I'm hoping this month is one big first step, and afterwards I'll have the strength to go on with it myself. Or not, and I'll move onto much harder drugs, and realise what a small problem alcohol really was.

The remainder of my Perth shows go just fine, with many sell outs. Worth celebrating with a drink, but I'm that desperate for one, it's not like I need an excuse.

Adelaide Fringe Festival, Days 45 to 49 – Friday, February 22 to Tuesday, February 26

Next, on to the Adelaide Fringe, and I'm seeing reasons to drink everywhere.

Right now it's Saturday, February 23, and so far, between the comedy circuit and mining site, I've been offered alcohol 237 times.

But who's counting?

Not me, I made that up. It feels like that many though, and if I were instead counting the number of times I've thought about alcohol, well it'd be in the thousands. Once you're craving it, you also see opportunities to drink everywhere. In terms of being offered drinks though, here's what happened today...

Chatting to some FIFOs at the hotel breakfast buffet, they suggest we go for a beer. It's not even 9 am.

Lunch with three other comedians. Who all have a beer.

At the supermarket in the bottle shop, there's a wine tasting.

After performing at an early afternoon children's show I'm offered a pint of cider, which it's difficult to refuse, even though I can't stand cider.

Following a spot on the Fringe Showcase Stage in the centre of a bustling Rundle Mall, I'm offered cask wine by a homeless person, something in a brown paper bag by a man in suit, and sherry by a middle-aged lady.

At a late-afternoon showcase gig, there's a bottomless jug of beer that's free for performers.

I meet a journalist for an interview, and am told I can have anything from the drinks menu.

At my venue, the bar staff are practicing making cocktails, and I'm asked if I'd like to try a few.

For about an hour, I hand out flyers dressed as a penguin – it's a marketing thing. More people offer to buy me a drink than a ticket to my show.

The venue manager gives me two drink tokens. When I attempt to purchase a juice, I'm immediately told that's an immense waste. So hand the tokens to a fellow performer and pay for the juice.

A group of four women come in late to my show, each holding a half-full bottle of wine. They sit in the front row, and attempt to fill my cup of water with wine, offer to sleep with me, then point out they're all old enough to be my mother.

Backstage at a late-night showcase is an Esky full of beer, cider and spirits. Then after my spot I'm at the bar, and a guy offers to buy me a drink. I accept and opt for a fire engine – lemonade and red cordial. He hands it to me and says, 'I thought you'd be shit, but you weren't shit, but next time someone buys you a drink, order a fucking real one.'

I really doubt that all this is normal, and I'm starting to think I must just look like the type. Or perhaps everyone remembers me from last year's Adelaide Fringe, back when I was drinking and of which, I don't remember much.

Dispatch No 25 – Wednesday, February 27 to Saturday, March 2

Aren't they amazing? As a kid, I used to shoot 'em

Day 50 – Wednesday, February 27

While on my two-week break, I'd called Jonno twice. No answer. So I assume I still have a job, as my flight back hasn't been cancelled, and you can't be fired while you're away. Isn't that a rule? Maybe Jonno forgot about that when he told me I might not be coming back, and I might get sacked as soon as I land. Or perhaps the rules are different for me, because I got the job through my uncle.

We land, and none of the buses heading off to the various campsites has my name on their list, and there's nobody waiting to collect me. I call Dale and Jonno. No answer. Maybe I've already been sacked, and they just forgot to let me know.

So I wait. For over an hour. Eventually, Dale phones. He's on his way.

Over the fifteen-minute trip to the job site in his ute, Dale tells me, 'Peter finally canned Jerome. Well, he made Jonno do it, and that poor prick Jerome had no idea it was coming. Just kept crying. He was told to go back to camp, but he just kept saying, "No. Let me finish out the day. There's still so much to do." As if the useless unit ever did anything anyway! It was actually quite funny, watching grown men trying to deal with this watery, snotty, all-over-the-place clown. He was properly hysterical, and it must be hard, being that fucking dumb.'

Jonno's in his office and stands as I enter, while Dale immediately goes for a smoke.

'You may as well set up in here, plenty of room,' he says.

'Okay.'

'Now how are you going with those user manuals?'

'I'm a fair way along,' I reply, although I haven't been able to bring myself to even look at them. 'It'd really help,' I continue, 'if you sent through some other examples from previous jobs. To give me an idea of what the finished product should look like, the type of language I should be using, that sort of thing.'

'I've already sent you through several examples,' he replies.

'I'm sorry, those weren't quite right. Could you send some others?'

He steps closer to me and bangs on the desk, causing me to jump.

'Xavier, mate, to be honest I don't know how they'd help. Nobody ever reads the stupid things. It's just a contract requirement we need to tick off to get paid. You want to be a writer, don't you? So just fucking write.'

'Got it.'

Jonno throws some car keys at me, although he's close enough to hand them over. They hit me in the chest, and I manage to catch them before they hit the floor.

'You'll need these,' he says. 'Your new campsite is an hour away in a car, or two hours on a bus. So the car's a piece of shit, and you can't drive it onsite. You'll have to park it at the security office, then walk if you can't get a lift. That's where it is now.'

'Okay,' I reply. 'If it's not mining standard, why is it even out here?'

'A bloke working for us on that Mackay Hospital clusterfuck borrowed it from a builder and backed it into an office tower. The halfwit. The choice was either pay $4,000 to fix it, or buy it for $2,000. So I bought it and drove it out here, and Jerome had been driving it, but it's yours now. Enjoy.'

Jerome had a car? His campsite was barely fifteen minutes from here.

The new accommodation isn't bad, but it isn't too good either, and after staying at the top notch campsite in town, it's horrid in comparison. As I'd been warned, dinner is a choice of steak, pasta or stew, which looks like last night's steak just in smaller pieces with more gravy. There are also vegetables and a couple of salads, and apparently it's the same every night.

My room is stark and depressing, and feels like an oversized prison cell, and it wouldn't surprise me if a huge tattooed man appeared and insisted that we share the bed. Really though, it's functional and fine, and I'm not about to complain, especially with my current level of job security.

Day 51 – Thursday, February 28

Over February, I kept a rough list of the reasons I craved a drink.

Now it's nearly over, here's a short summary:

To celebrate leaving the mining site for two weeks.

To celebrate arriving back at the mining site.

Waiting for a friend.

When a friend texts to say they're running late.
Catching up with friends.
Girls.
Talking to them.
Watching them (in a non-creepy way).
Wanting to talk to them after watching them.
Talking to them and not knowing what to say that isn't extremely polite, because I feel bad after spending so long watching them.
Getting dumped by them.
Because, in a bar, it's what you do.
Homeless people.
Being asked while flyering, 'Can you tell us a joke?'
Being told while flyering, 'That's not funny.'
Seeing a whole lot of my flyers on the ground. In the bin. In the urinal.
Anytime I was wearing that now very sweaty and stained penguin suit.
The guy with a disability on the bus, who I wanted to help, but didn't.
Feeling guilty for being so able bodied, and still whingeing about so much.
Before and during every single stand-up performance.
After a stand-up gig goes well.
After a stand-up gig goes poorly.
After a stand-up gig goes just okay, and I can't figure out why.
Being told how funny I am. Being told how funny I'm not.
Poor ticket sales. Great ticket sales. Average ticket sales.
Realising how little there is left over after expenses and the box office takes their cut.
For an extra kick, to stay up late.
To quieten down a racing mind when it's time for sleep.
After every awkward experience.
After an argument.
To fill in time.
Because I know that if I have a drink, I'll be able to stop thinking, relax and feel good about myself.

Day 52 – Friday, March 1

February is officially over, I can drink again and it's a RDO. How perfect is that?

From the moment I wake up, I'm acutely aware that I could have my first drink at any moment, but there's no rush. I know that at some stage, it'll happen. It's as if I'm deciding when to move in and kiss someone for the first time, and I'm already certain that they're going to kiss me back.

On my last night in Adelaide a comedian asked me, 'Are you going to have a drink on March 1st? Or are you going to keep this non-drinking thing going?'

There was never any doubt – I was always going to get back on the booze. Back in February with just a week left I knew I was going to make it, but I still got extremely jittery, and constantly fantasised about sneaking a sip from some of my favourite illicit drinks. A beer in the shower, hip-flask while on a train, afternoons in city parks with longnecks in brown paper bags, filling my water bottle with white wine for a music festival, sneaking shots of whiskey into my midday coffee.

Over the final few days of February, the team at Febfast sent out several emails warning people to take it easy on March 1st. To be aware that after a month off alcohol your tolerance would be lower, so you should drink plenty of water, take breaks, and eat plenty if planning to hit it hard.

But where's the fun in that?

The closest pub to my campsite is the Jolly Collier Hotel, where I head at midday after eating and drinking nothing all day, and I'm so excited that I attempt to down all of my first beer in one go. Halfway through I start to feel dizzy and am hit by a piercing ice-cream headache, but I still empty the glass, and apart from my throbbing head, the alcohol's already wrapped the rest of me in a blanket of gently buzzing warmth.

After two more quick beers I spot some FIFOs, and for a few minutes, it's the best conversation I've ever had onsite. We're laughing, shouting, patting each other on the back and I'm coasting, sailing, and then airborne I'm so happy, until I realise that it's all fake. My laughs are forced and louder than their stories deserve. While everything I'm saying, I'm inventing on the spot by mixing together information they've given me, and whatever random rubbish is in my mind at that moment. Not that it matters, they're not even listening. Their eyes are glazed, and they're swaying, slurring, and downing rum and cola as if the stuff is air, and just as cheap.

Around 2 pm, I declare that I'm leaving. The FIFOs implore me to stay like it actually matters to them, but the moment I step away from their table, it's as though they instantly forget that I ever existed.

Back at the campsite I churn through the comedy admin and start worrying. Now that February's over, will my efforts to abstain actually make a lasting difference? I've taken a month off alcohol before, then on March 1st embarked on a three-day bender which resulted in me misplacing an iPad, mobile phone, mountain bike, rain jacket and girlfriend.

That time, I stopped drinking to prove to the girl that I could. This time I did it for me, as well as a girl, and I've definitely learnt some different things.

By repeatedly denying myself a drink when I thought I needed one, I discovered something that should've been obvious. I never need one, I drink because I choose to.

Also, consuming copious amounts of alcohol is something that's revered in Australia.

When you repeatedly hear things like, 'Far out. He goes HARD. Classic Xavier,' well it doesn't feel like you've got a problem. It feels like you've just successfully run a marathon in a desert up a steep mountain.

We're a society who seems to celebrate everything with alcohol, and when you're not drinking, it causes problems. Unless you're pregnant, driving, or a recovering alcoholic.

'Why wouldn't you have a drink? Aren't you having a good time?'

'Can't handle your grog, mate?'

'What's wrong with you? Are you soft?'

One surprising thing I discovered during February was the number of people who told me they couldn't do it, and admitted they had a problem. They weren't slurring, stumbling, swaying or covered in filth, and I'd never realised I knew so many people that had issues with alcohol. This only happened, however, while I was in Perth and Adelaide for that two weeks. Nobody at the mining site ever once suggested that they were considering drinking less.

Febfast was also about the fundraising, and I should've pushed that harder. I never mentioned it to anyone onsite, and half the point of joining Febfast was to spread the message, while the other half is raising money and, out here, there's plenty around.

I've also learned first hand that most pubs, clubs, restaurants and cafes shut off their coffee machines several hours before they close, and offer very limited other options. It'd make being in these places and not drinking so much easier if more of them offered something other than alcohol that was worth drinking.

From others who've spent time off alcohol in the past, I heard that while abstaining they had more energy and got much more done. Personally, I noticed a slight improvement but nothing revolutionary. For me, the main change was my mood. There were still good and bad days, but no run of days where it was all too much and there was no point to anything, which have recently seemed to follow each big night out.

Also I was terrified of failing at comedy while sober, and the key lesson I learnt from a month off alcohol was how to manage that fear. Whenever I'd failed when drunk, at least then I had alcohol as an excuse. There's also respect to be found in drinking as much as you can and still being able to function, but I want to make comedy a career, not something that I need a separate career to pay for. So no more heavy drinking prior to performing.

Outside of comedy, whenever I'm in a situation where others are drinking and I'm not, I still desperately want to and struggle to socialise. During February I solved this by avoiding late nights, but I'd like to be one of those people that can have a few drinks on a night out without needing to get drunk, and the last thing I learnt is that I'm not there yet.

That night I call Verity. She answers, which is very exciting.

'You made it?' she says. 'I am very impressed. Well done.'

'Thanks. What's news with you?' I ask.

'Brendan and I are moving in together.'

I want to ask, 'Who's Brendan?' or 'Isn't that kind of fast?' but instead I say, 'That's great'.

Day 53 – Saturday, March 2

It's a ninety-minute drive between my new campsite and the job site. Jonno previously told me it was an hour, but I think that's only if you break a land-speed record, or have a helicopter. On the narrow and winding roads I am passed every morning and evening, however, by an intermittent stream of trucks, utes and cars who seem intent on breaking that record, especially around blind corners.

Up and in the car by 5 am, then returning every evening around 7 pm, I am treated to each day's sunrise and sunset. Sometimes it's too cloudy to see anything, as it's still the wet season, but most days it's clear and I get the full show, from star-filled black, through navy blue, purple, red, orange and yellow, then on the way home, the same show in reverse. Any solitary clouds have their edges lit golden, and if thin enough, their interiors ignited a muted version of the sky's colour in that moment.

On colder mornings there's a mist that hovers over the road, gathers around trees and hides everything beyond. During my drive it's gradually burnt away by the heat of the day, unveiling glowing orbs to be cars, and fields, factories, farms and mines are all gradually revealed, as if the focus dial on a telescope is being slowly turned.

As I'm one of the few travelling from that particular campsite to the mining site where I'm based, there's a short stretch of road that twice a day, I seem to share only with a huge flock of brown whistling kites – a bird of prey found throughout mainland Australia, their wingspans up to one and a half metres. Always in this same area, sometimes there's over a hundred of them in the sporadically spaced trees lining the road, or flying just above, or standing on the road. Considering their impressive size, it seems to be a more arduous and longer process for them to take flight than smaller birds, so I always slow down when they're around. Despite many near misses I've never hit one, and every morning so far of doing this drive I've noted that I'm yet to see a squashed one.

Dale calls this morning, just as I'm leaving.

'Xavier, I need you to pick me up on your way through town. I got a lift in to pick up some smokes, but can't get a lift back.'

'No worries,' I reply.

'How far away are you?' he asks.

'An hour and a half.'

'That drive only takes an hour for a normal person. Hurry the fuck up.'

It's after 6.30 am when I finally arrive, and I've never seen Dale so angry.

'Settle down mate,' I tell him. 'It's only half an hour.'

'Fine for cunts like you who do fuck all.'

We drive past a small flock of kites.

'Amazing birds,' says Dale. 'As a kid, I used shoot 'em. For target practice, and 'cos it's fun.'

After parking at the security gate, Dale rings Ben to come and collect us.

'Sorry for blowing up at you earlier,' he says to me. 'I've been out here nearly 50 days straight now, and I've just had it, ya know?'

'Fifty days? I thought the maximum was five weeks?'

He scratches at the back of his head, like he's trying to remove something. 'The maximum is whatever it takes to get the job done. Some days, I just want to punch cunts for no reason. Happens to everyone after too long out here.'

'Can't you tell them you need a break?' I ask.

'If I don't do the work, they'll find someone who will and I need the money. Wife, house, kids, school, one failed business, the regular shit.' He loudly clears his throat, and it turns into a violent cough. 'I'm sorry, I shouldn't even be talking to you about this.'

We get out of the car, I lock both doors and Dale laughs.

'You're keen. Thinking someone would bother stealing that thing.'

'If you ever want to have a chat about any of this, let me know. Or maybe talk to someone?'

'What like a head shrinker?'

'Why not? If you're feeling depressed or a bit sad or anything.'

'I'm not fucking depressed. Fucking feelings are for faggots.'

Ben arrives, they both light up smokes and Ben drives slowly across the site, so they've got time to finish them before we arrive.

Dispatch No 26 – Sunday, March 3 to Wednesday, March 13

There's nothing happening

Day 54 to Day 64

Nothing happened.

Dispatch No 27 – Thursday, March 14 to Friday, March 15

Money is all that matters

Day 65 – Thursday, March 14

JRT Projects' two bosses, Peter and Scott, arrive today. Most of the job has now been completed so they're here to meet with Debitel about payment, and it could very easily get very tricky.

Basically, the building industry is based on deceiving people and screwing them over.

Out at this job site Debitel is the head contractor. It won the contract to build this mining camp, and is paid by the mining company through Nuscon, the project managers.

Now Debitel won the contract through a tendering process, where a whole bunch of companies estimate how much the job is going to cost to build, and the one who comes in with the lowest number wins, provided whoever's in charge believes they can actually do the work to an acceptable standard. Sort of like the way you might not choose to buy the cheapest television you can find, because you'd like one that won't spontaneously combust, and burn down your house while you're sleeping.

So the whole tendering process is often incredibly corrupt. Inside information is shared, backhanders are offered and deals are done, however, it's all kept relatively honest by the simple truth that once it's all wrapped up, the job needs to be profitable enough for all the companies involved to stay afloat. Which doesn't always happen.

Currently, the building industry in Australia is ridiculously competitive and often a contractor will quote a price that's very close to or even below what they expect the job will cost them to build. The actual building, however, never goes to plan. It always includes variations, and this is where there's money to be made. Since the additional piece of work isn't in the original contract, the builder gets to pretty much charge what they like, without being too extravagant.

There are always clauses in the contract that are meant to limit the builders' ability to make an outlandish profit on these variations, but they're often not very effective.

Now JRT Projects is working for Debitel as a subcontractor, and we're claiming a lot of variations – most of which we're still waiting to be paid for. It's a delicate process, as Debitel has to get the money from Nuscon, who has to apply to the mining company for it, and they only release the money if the work has been done to their satisfaction. Meaning that the whole process often takes months.

The biggest problem is that any job to do with mining costs millions to complete, which the different contractors and subcontractors have to pay out by way of materials and labour, then might not see back from the mining company for months or sometimes years, the whole time keeping their fingers crossed and hoping not to go bust in the meantime.

I have no idea what the financial situation is at JRT Projects, but I do know a few suppliers haven't been paid for months, and both bosses wouldn't be here if there was nothing to worry about.

So the big meeting is tomorrow, and Dale and Jonno did check the variations I put together, but I have no idea how closely, and if Debitel find any holes the whole thing could fall over.

Day 66 – Friday, March 15

During the meeting between the Debitel boss and four of us from JRT Projects, the imposing, pristine folders of variations that I've spent months assembling are not even opened before one major point of contention brings the whole thing to an end.

'That stuff in there,' the Debitel boss nods at the folders. 'Have you got work orders for all of it?'

Peter looks at Scott who looks at Jonno who looks at me, and when I look back at Jonno, he says, 'You know a lot of that stuff was done based on a handshake.'

'All I know is what's stated in the contract. Which is that no work is to proceed without a work order,' says the Debitel boss.

'What the fuck are you on about?' Peter asks him.

'He's saying that he's only going to the pay for the stuff that's been properly signed off. Have I got that right?' says Scott.

'Correct,' says the Debitel.

Peter turns to Jonno. 'So how much is that? All of it? Half of it? None of it?'

Jonno looks at the wall. 'Not sure.'

'Well, unless you can produce some evidence, we won't be paying you anything,' says the Debitel.

'Just take a look out of that fucking window, mate!' Peter

stands and points outside. 'The work's done, and now you're not fucking paying us for it?'

'Just the variations,' the Debitel replies.

Peter, Scott and Jonno are silent, and my guess is that those variations are the difference between JRT Projects making a profit or a loss.

After the meeting and back in the JRT onsite office used by the tradesmen, Peter is emptying the contents of filling cabinets onto the filthy floor. 'So does someone want to tell me what I'm looking for?' he asks.

Donk shrugs and keeps playing solitaire.

Jonno then enters and tells Peter, 'Settle the fuck down, will ya? Xavier and I have got this.'

'You fucking better,' says Peter.He then looks at Scott, who's spinning himself back and forth in an office chair.

'Pub?' Peter asks, and Scott nods.

Once they've left, Jonno and I spend the night collating any work orders we do have, and where we can't match one to a variation, we dig through emails, scrawled notes and text messages for whatever evidence we can find.

Just after 4 am we've found enough evidence, however flimsy, to back up every variation. What we don't have is any proof of the outrageous costs we've claimed each variation cost to complete. Luckily nobody's picked up on of that, because that justification simply doesn't exist.

'Don't know about you,' says Jonno. 'But I'm going back for a quick shower and some food. I need to be outside of this room for at least an hour. It really does smell like the inside of a diseased anus.'

'We probably should've done all this from the other office.'

'And you probably should've suggested that about eight hours ago,' he replies.

Day 67 – Saturday, March 16

At today's meeting are the same people as yesterday, with the only difference being that there's an additional thick folder on the floor.

Debitel's boss nods at the new folder. 'So that's the evidence is it?'

‘Correct,’ says Jonno. ‘Take a look if you like.’

‘No thank you.’ He turns to Peter and Scott. ‘So I’ll pay 70 per cent of what you’re asking for. Or we can go through every one of those variations and you might see some money a year from now.’

‘Or we can take you to arbitration and get the full 100 per cent, along with some hardship costs because you’re being a prick,’ says Jonno.

Peter and Scott stare at each other for a few seconds, and Scott gives him an barely perceivable nod.

‘We’ll take the 70 per cent,’ says Peter.

Later that night, Jonno catches me in the dining room.

‘Some fine work there, putting those variations together,’ he tells me. ‘Now I can’t work out if Peter and Scott needed the money, or they’re just weak pricks, but I know that if it went to arbitration we would’ve fucking won easy. I told Peter that before the meeting, because I thought it might go that way. Arbitration is all about your evidence up against theirs, and I don’t think Debitel’s got much of any, as they’ve been caught up with trying to get money out of Nuscon and the mining company.’

‘Really? No evidence at all?’ I reply.

‘Well, we’re never going to know now are we? So I suppose none of it fucking matters.’

Dispatch No 28 – Sunday, March 17 to Friday, March 22

There's still nothing happening

Out here, things that should take one day often take six. At least. Also, it's very much a case of the same shit day after day. After day. Here's a snapshot.

Day 68 – Sunday, March 17

Eighteen days since my last break. I wake up at 4 am, an hour before my alarm, and I should go to the gym, but instead I watch the end of a poker tournament, ten minutes of two different movies, and an infomercial featuring an exercise contraption that looks like a pink plastic spider that's trying to wiggle the limbs off a woman with 6 per cent body fat, who's apparently had three kids.

I should get on to those operation manuals, but I really, really don't wanna. Instead, for my first three hours at work I do comedy admin, which today involves compiling a list of media contacts. I need to get all the admin out of the way, so I can get onto writing some jokes and have something worth saying when I'm onstage in Melbourne.

Just before morning tea, Ben bursts into the office.

'Xavier, I'm done. I need to quit this shit. Is there a form or some crap I have to fill out?'

'Don't think so. Just tell 'em after this swing that you're not coming back.'

'But is there some way I could put it in writing? To make it official and all that?'

'You could do a resignation letter.'

'So if I bring one in tomorrow, you'll look over it for me? For the spelling and shit?' he asks.

'Does it matter? I mean, if you're quitting anyway?'

'I don't want to fuck it up and still have a job.'

'Fair enough,' I reply.

Ben leaves, and he's at least the tenth JRT Projects employee since Xmas to threaten resigning, and most of those have done so multiple times. None have followed through.

Roy from Nuscon, the project management company, has emailed Jonno, my boss, requesting some pipe-testing documentation and videos, and Jonno asks me to find them. It's something else Jerome was supposed to have done and he's had a go, but the videos, photos and documents have been stored so illogically it's as if he was trying to hide them, and I need to start over.

Now the mining camp we're building isn't properly finished, as we're still here working on it every day, but it's apparently finished enough because FIFO workers have started moving in, and today it's my turn.

So I move into another donga, just like any other, but it's the largest I've ever seen and although it's new it's hard to tell, as everything's been done in the timeless cheap decor of a budget hotel room, lacking only the horrid carpet and artwork.

The bright, white dining room also feels temporary and timeless, like it could've been here for an hour or twenty years. Inside are swipe card turnstiles, tables, chairs, cutlery and crockery, as well as bain-maries and salad bars stocked full of food and condiments.

While eating alone, I overhear several complaints about the possible alcohol restriction. On all mining campsites it's apparently a rule that you're only allowed three mid-strength cans a night, however, I've never met anyone who's stayed anywhere it's been enforced.

Back in my donga it's bedtime and the hot water doesn't work, and neither does the air-conditioner or the brakes on the wheels of my single bed, so while I fall asleep near the far wall, I wake up blocking the door.

Random highlight of the day

This conversation was overheard in the dining room between two DIDO workers who'd just returned for their swing.

'Fell asleep again,' says DIDO one.

'Me too! How long into the drive were you?' asks DIDO two.

'Like two hours. Of seven.'

'Haha, you fuckwit! So how the hell did you make it?'

'Certainly perks you up a bit, when you wake up in a ditch,' replies DIDO one.

'Car okay?'

'Fine. Don't put the cruise control on anymore, so the car just rolled off the side. All gentle like. Crossed a lane though.'

'That's fucking dangerous man,' says DIDO two.

'Out here, what are you going to hit?'

'Trees. Other people. Plenty of shit.'

'Didn't you just do the same thing?' asks DIDO one.

'I just nodded off for a bit in the servo carpark.'

'Bullshit.'

'Maybe. Hey, this dessert is fucking delicious,' says DIDO two.

'Is that soy sauce on your cake?' asks DIDO one.

'Fuck oath it is. Proper sweet and sour right here. Fucking taste sensation.'

Day 69 – Monday, March 18

One big plus of living where you work is that my commute has been cut down from a ninety-minute drive to a ten minute walk. At this morning's site-wide safety briefing a few of the guys complain that it's too far and request a shuttle bus.

Leon, the head of health and safety onsite, replies, 'No way am I getting a mini-bus just for you fat pricks who can't be fucked with a short walk.'

The company I work for, JRT Projects, was responsible for installing my hot water unit, so I ask Dale about getting it fixed. For the rest of the day he tells anyone who comes into the office, they all laugh, and I wonder if it's a planned prank.

After scanning through the long list of items that need to be included in the manuals, I spend five minutes searching for them, then give up and send off a barrage of emails, knowing that if I wanted results, I'd follow up with a phone call. Really though, why do something yourself, when you can put it off by waiting for someone else to get back to you? With the comedy stuff, I send out my press release to media contacts, and this takes up most of the day.

Nineteen days since my last break. Nine days until my next one.

Every time the office door has opened today, I've been listening out hoping to hear the voice of a friend from home. Although there's no way that could ever happen, pretending it might tricks my mind into believing that it's just about to and boosts my mood.

Early afternoon, Ben says, 'I'll bring in that letter tomorrow. Will you have time to have a look at it?'

'Sure.'

Damo has some of the pipe-testing videos that I'm still

missing on his phone, which I download onto my computer, burn to a DVD and hand-deliver to Roy. When I return, Damo, Ben, Jonno and Dale are at my computer, staring at my Melbourne International Comedy Festival press release. I'm as embarrassed as Donk should be each time he's caught looking at porn, which he seems to have given up on, as each time I've glanced at his screen lately he's been playing solitaire.

'What's this?' says Damo. 'Is it supposed to be funny? Because it's fucking not.'

'Show us one of the videos you got off Damo's phone, will you?' Jonno asks.

I start playing it.

'Is there any sound?' asks Dale.

There is, but it's currently running through headphones, something else you can get instantly dismissed for using onsite, and the moment I unplug them from my laptop Damo and Pando can be heard shouting over the entire four-minute test, with a constant and clear stream of 'fucking' this and 'cunting' that.

Knowing that it's already with Nuscon and their offices full of safety and human resources personnel, it's as if I'm suddenly hearing the video with different ears, and I imagine those at Nuscon who aren't currently filling out complaint forms have fainted, because while Roy and the other Nuscons swear plenty onsite, I've spent enough time in the offices of similar companies to know that once they're back among their own, that sort of language isn't tolerated.

'Are all the videos like that?' asks Jonno.

Damo has gone bright red. 'Yes.'

'You've already delivered them to Nuscon?' Jonno asks me.

I nod.

Dale and Jonno both laugh.

'Get them back, so we can cut it out,' says Damo.

'Don't worry about it, they probably won't ever watch the fucking things,' says Jonno.

'And if they do,' says Dale, still laughing. 'Who gives a fuck?'

Then just after 4 pm Dale explains that they need a work order before anything can be done about my hot water. So I lodge a request with the mining camp administration, who move me into a different donga where the air-conditioner, television and hot water don't work, so I immediately return to my previous room.

During the day we were all given swipe cards for the dining room, and were told security would be barring anyone without one. That evening, none of the cards work so we all walk around the turnstiles, while two security guards watch in silence. Earlier in the day, a note was slipped under the door of every donga detailing the rules of the alcohol restriction, and over dinner the guys threaten to quit, walk offsite or go on strike. Just the same as when there was a rumour floating around that smoking would be completely banned.

After a cold shower I walk one lap of the campsite, which takes twenty minutes, and get a few strange glances from the groups drinking and smoking. Everything is so neat, exact, simple and similar, with large blocks of unfinished construction everywhere, especially at the edges, it's as if I'm trapped in a half-finished jigsaw puzzle.

Random highlight of the day

Yesterday a delivery driver tried to drop off a fridge that nobody had ordered. According to Donk, 'This prick had pupils the size of dinner plates. I swear his eyes were spinning around in his head, and he smelled like he'd pissed himself.'

Today, when a different driver arrives with the same fridge, I'm the only one in the office.

'If you're not going to let me leave it here, where the fuck should I leave it?' he asks.

'Sorry mate, I don't know.'

'I do deliveries. I've got to deliver it.'

He's holding a yellow slip.

'Give me a look at that?' I ask.

It clearly states the shipping address as a mining site about an hour away, and I point this out to him.

'But that's not the delivery address,' he replies. 'And the pieces of paper are usually white.'

'I really think that shipping address is where it's meant to go.'

'Don't be a smartarse,' he tells me. 'I couldn't take it there even if I wanted to, because I haven't got a fucking consignment number.'

'So how do you get one of those?' I ask.

'From you I think.'

'How did you even end up here?'

'This is the last place we delivered a fridge,' he replies.
'I'm really sorry, I can't help you.'
'So who can?'
'I don't know mate. Maybe call your office or something?'
He breathes out and uses his mobile phone to make a call. As he leaves the office I hear him shout into the phone, 'Sorry boss, this JRT bloke is a right fucking wanker. Just like you said he would be.'

Day 70 – Tuesday, March 19

Twenty days since my last break. Around 12 × 20 = 240 hours at work. I can't wait to wake up and not see the inside of my donga. To eat something that isn't always in the same place, from an identical selection, put there by the same miserable person.
No replies to yesterday's emails, from either comedy media people or about the mining manual stuff. So if Jonno asks how I'm going, at least I've done something, but I also remember him telling me that nobody ever reads the manuals and to just get them done. So if he pushes the issue, I'll be easily found out.
Ben drops off a small stack of DVDs.
'These are the movies or whatever from inside the pipes. Look out for the frog,' he says, and leaves.
Donk immediately crosses the office and demands to see the frog video, assuming that it's a sex thing. It's not.
In order to check each pipe's integrity, a camera is fed down its entire length, and in one of the videos the camera disturbs a frog. Which proceeds to leap down the pipe just ahead of the camera for the entire twenty minute duration of the test.
Although everyone is supposed to be outside working, a cheer squad of eight quickly gathers for the frog around my laptop. A few times the camera catches right up to it, and the frog looks to be flagging and nearly done, but it always rallies, almost as if the cheers from the guys push it along. Then there's a light at the end pipe, the frog speeds up, and the video cuts out.
The guys turn to me. 'Where? What? Make the frog come back. What happened to it?'
I shrug.
'That's a sewer pipe, which leads to the sewer water treatment plant. So it's probably dead,' says Dale.

There are several frowns and scrunched up faces, and I'm worried that Roger or myself might shed a tear.

'However!' Dale continues. 'The treatment plant isn't online yet, and the emergency overflow goes into the creek.'

'So it's okay?' asks Roger.

Dale laughs. 'No, you idiot, of course it's fucking dead. Drowned by all our shit.'

A request comes through from the mining company to quote the cost of fixing my hot water unit. So with Dale's help I put together the variation, and including labour and materials it's over $3,000.

Ben laughs loudly. 'Give me $50 and I'll do it right now.'

'Seriously?' I ask.

'Well, in the real world I would, but this isn't the real world.'

Jonno later tells me that since this mining camp has opened, we've already put in claims for over $100,000 worth of additional variations on top of what we've already claimed.

At 4.45 pm, as the JRT employees file in to drop off their safety paperwork before leaving for the day, Ben whispers to me as he passes, 'I'm definitely doing that letter tonight.'

Which reminds me of something the uncle who got me the job out here always says about his years as a nightclub bouncer. 'As soon as they start talking about hitting you, they're never going to hit you, because if they were really going to throw one, they would've done it already and wouldn't be going on about it. The proper crazy guys, they just do it without thinking.'

Tonight at dinner the turnstiles are in operation, and everyone's card works. Unlike the previous camps I've stayed, at this campsite there are people I know everywhere, so I feel obligated to make an effort. So far my pleasantries have been met with curt replies or ignored, and I'm surprised to feel more sad than relieved.

At least I've jammed enough cardboard under the wheels of my bed that it moves only inches this evening, instead of metres.

Random highlight of the day

I like the idea of being outside the office, but actually being outside is awful. It's so hot that every limb feels heavy, and sticky, and anyone I see always asks me, 'What are you doing out here?' or 'What's wrong? What are you looking for?'

Standing well behind a barrier of orange flagging that's ignored by most, I observe as one of the last pieces of intense construction continues. Diggers, bobcats, cranes, dirt movers and workers zip around drilling, connecting, carrying and filling in large rectangular holes of varying depths, with pipes and conduits packed full of wires poking up through the dirt.

On adjacent concrete slabs are large tanks and electrical panels, with different sized pipes and valves zig-zagging throughout, and workers meandering about doing the final checking.

Nearby, mechanical arms connected to drills and buckets move as quickly and nimbly as the human arms directing them, scooting around and depositing piles of gravel, carefully dislodging, breaking apart and removing boulders, levelling out sewer trenches at the correct gradient, and delicately nudging pipework and tanks into the proper position.

Day 71 – Wednesday, March 20

As I'm leaving at 6 am, a guy arrives to look at my still non-responsive air-conditioner. He opens the remote, turns the batteries around, closes it and turns it on. Well, that was embarrassing.

At breakfast the turnstiles are working but I can't find my card, so I head across to administration for a replacement. Which can only be issued after I fill out a form that's verified, sent off somewhere else, and then approved. I'm told to check back that evening, and there are supposed to be single-use meal cards but they haven't arrived yet, and nothing else can be done as all their phones and internet have ceased working for some unknown reason.

First thing at work this morning, Ben's at my desk. 'Can you email me an example or something of what this letter thing is meant to look like?'

'All it has to say is that you resign, and the date you want your last day. You're on a casual contract like everyone else, aren't you?'

'That's right.'

'So they can fire you whenever, but you can quit whenever you like. You don't need a letter.'

He nods. 'You're right, but email me something anyway, okay?'

So I immediately search for and find a few resignation letter templates, then email them through to him.

Two companies have sent through up-to-date product instruction manuals, so I create a folder on my desktop and add them to it. To everyone else, including the comedy contacts, I send a follow up email. Then log onto the JRT Projects' server, find some operation manuals from similar jobs, and I add them to that folder on my desktop – meaning I'm now even more ready than ever to get started.

The variation to fix my hot water has been approved, so Dale gets Ben to do it, and he's back less than half an hour later.

Roy sends another urgent email asking Jonno for different videos and photos, because apparently those I sent through weren't what they were after. Roy doesn't specify what he does need and won't answer his phone, so Jonno asks me to collect and send everything I can find before I leave for the day. Last time I only went back two months, but now I have to look through 18 months of files, meaning I'm at work until 7 pm sorting through hard drives, memory sticks, DVDs and cameras on a tedious treasure hunt for anything useful.

The videos I find are all work related, but only a fraction of the photos feature the work we've done with the date spray painted onto the ground as proof. Instead, the pictures are of hairy, muddy and unsuspecting arse cracks, guys pretending to ride machinery or pipework rodeo style, guys asleep at work, or guys after work surrounded by empty beer cans, on nudie runs, or bent over fresh piles of vomit. Scrolling through hundreds of similar photos, I have no idea what compelled these men to record and keep evidence that could get them instantly fired.

From a list of all the testing videos that I'm supposed to have I notice that a few are missing, so I just randomly duplicate a few that I do have, then change the dates so it looks like there's a complete set.

So I haven't had breakfast, or lunch, and still don't have any meal passes, so no dinner either. Luckily most of the men take so much unnecessary food into work every day that from the JRT office fridge I'm able to scavenge more than enough for several meals.

While drinking a mid-strength can outside my donga later that evening, I see others sipping from plastic cups as part of an effort to disguise their full-strength beers. Then I overhear that nobody's stopped making runs into town for beer and spirits, and despite the threats, no vehicles have been searched.

Also, I still don't have any hot water.

Twenty-one days since my last break. Seventy-one days since my last night with Verity. Or anyone.

Random highlight of the day

This conversation happened at 10 am over the radio, and was between the onsite manager from Debitel, and the visiting international head of human resources from Nuscon.

'What happened to Hoopy?' asks the Nuscon.

'That useless cunt? I sacked him,' replies the Debitel.

'Why?'

'Because he was a useless cunt,' says the Debitel. 'Like I just said. Fuck, are your ears broken, or are you just a dumb cunt?'

Day 72 – Thursday, March 21

Administration is still clueless about my replacement meal card, so at breakfast, I deftly sidestep the turnstile and the security guards do nothing.

At the JRT morning briefing, Jonno says, 'Everyone's moved in here now, correct?'

Nods all round.

'Anyone got any problems? For example, everyone got hot water?'

Silence.

Jonno looks at me. 'Well who wants to have a hot shower with Xavier? This poor prick will do anything for one, the amount he's been whining about it.'

Everyone laughs, some so hard they are literally crying and slapping their knees.

Before I've even made myself an instant coffee this morning, I click open the folder of manuals I've compiled. Then close it, and spend the time until lunch on my Melbourne International Comedy Festival promotion – setting up Facebook events for every show, and adding it to event listings and 'what's on' websites.

Dale then tells me my hot water will be fixed today. Since one variation has gone through it's now under warranty, so there's no point putting through a request as JRT can't make any more money out of it.

Twenty-two days since my last break.

From the smoking area beside my office I can see a cow paddock, and today I count four cows. There's usually six. Then I put salt in my coffee, because I've never done that before. Next I add pepper, and it's horrible, but instant coffee's always horrible.

After lunch, Roy comes into the office and I ask if he got all the stuff he was after.

'What stuff?' he asks.

'All that testing documentation and evidence.'

He stares at me, then hands me a DVD. 'Give this to Dale. It's the horse's most recent race. Finished fourth, but it's a good fourth. Oh, and show me that frog video?'

That evening, I stare at the security guards while bypassing the barrier. One stares back blankly, while the other smiles.

After dinner I sit down next to Ben, who's drinking alone outside his donga.

'I'm going to stick it out,' he tells me.

Random highlight of the day

There's a gym at this new campsite, as there is at every campsite, and tonight one of the two short, blonde and muscular personal trainers is in her exercise clothes and handing out free mid-strength beers, in an attempt to coax guys in for a free introductory session.

'I'm not a personal trainer,' she corrects guy one. 'I'm the lifestyle coordinator.'

'So what's the difference?' asks guy one.

'It's got a different name. Aren't you listening, you dickhead?' says guy two. He looks at gym girl. 'Sorry for swearing.'

'It's not just about training,' says gym girl. 'I can help with diet, exercise, your whole lifestyle.'

'Got your work cut out for you there,' says guy three.

'What sort of message are you sending, do you think, by handing out free grog?' asks guy one.

'Everything in moderation,' replies gym girl.

'Tomorrow morning, what time does it open?' asks guy two.

'We're not open yet.'

‘So you’re here handing out free beer, to encourage us to exercise at a gym we can’t go to, even if we wanted to?’ asks guy three.

‘It’ll be open in a week or so. We hope.’

‘Will you be handing out more free beer?’ asks guy two.

She smiles, ‘Probably not.’

‘You’re going to struggle for numbers then,’ says guy three.

‘We’ll see about that.’

‘One thing you won’t see is the three of us anywhere near the joint,’ says guy one.

Guy three grabs at his large stomach and gives it a wobble. ‘This took me years of no work. I’m not about to waste all that.’

Gym girl’s wide smile has been fixed in the exact same position on her face for at least five minutes. ‘Regular exercise will help you live longer,’ she says.

‘I’m here for a good time not a long time,’ says guy three.

‘These fucking mid-strength beers are only $2 each. It’s not like we can’t fucking afford our own,’ says guy one, who then tells gym girl, ‘Sorry for swearing. Again.’

Guy three reaches into a small Esky, gets out three heavy beers and hands them around, then offers one to gym girl. She declines, and guys one to three drop their barely touched mid-strength cans into a rubbish bin, and each lands with a dull, heavy thud.

Gym girl moves across to a different group of guys, where she points out one’s glasses. ‘They’re interesting safety glasses,’ she says.

‘No they’re not,’ he replies. ‘They’re just glasses. You know, all the better for seeing you with.’

Day 73 – Friday, March 22

Today, my hot water works, so I thank Dale for getting it fixed, but apparently they were too busy yesterday and nobody got a chance.

‘Do you still want someone to have a look?’ he asks.

‘Don’t worry about it,’ I tell him.

Twenty-three days since my last break. Five days until I fly out. It’s back to six cows.

After cruising news and sports websites for a couple of hours, I write a list of everything comedy and work wise that I still need to get done. It's almost identical to the list I put together a week ago.

Roy requests more testing reports, quality assurance and safety documentation. I send it off. He then requests more and I send him more, some of which I have to fill in myself, because it was never filled out at the time. In total, it's over a thousand pages.

That afternoon, Roy appears. 'Dale around?'

'Haven't seen him,' I reply.

'Get him to give me a call will ya?'

'No problem. All that safety stuff, I've been sending it through – it's what you're after?'

'No idea,' he replies.

'What do you mean?'

'People higher up ask me for it. So I ask you for it, and once you send it, I send it on. I've got no idea what the fuck it's actually for.'

It's about an hour until the end of the workday, and the office door slams into the wall, causing both to rattle, and I'm shocked out my half-doze.

Ben sees that there's nobody else in the office and says, 'I've fucking had it with this shit.'

For ten minutes he shouts about all the reasons he's going to quit, but never mentions any dates or a resignation letter.

Every day out here, if you listen closely enough, you'll hear someone saying very similar things.

At dinner the turnstiles are back to broken, around the camp nobody's bothering with plastic cups anymore, and I notice several utes come into the camp and past security carrying clearly visible cases of full-strength beer and pre-mixed spirits.

Random lowlight of the week

While lining up for dinner, I turn around to Dale and say, 'I just think this whole turnstile rubbish is such a waste of time and money.'

From where he's sitting a few metres away, Pando interrupts by shouting, 'Shut the fuck up Xavier. I've told you before, nobody cares what you think.'

Everyone laughs.

I can't wait to get out of here.

Dispatch No 29 – Saturday, March 23 to Monday, March 25

What's the grog and cigs got to do with anything?

Day 74 – Saturday, March 23

Just before lunch Leon comes into our office and informs us that earlier today, a bus driver died.

After a long silence, Dale is first to speak. 'That's fucking horrible. After work tonight, how the hell are we meant to get home?'

There are a couple of smiles.

'If there's a spare ute,' says Donk, 'maybe we can get Xavier to drive it?'

A few laughs.

'We should get the dead bus driver to do it instead. That'd be a whole bunch safer,' says Ben.

Leon looks at me and says, 'Beep, beep, beep!'

This time, everyone laughs.

Leon explains that, 'One of the two ambulances is in for maintenance, the bus driver's in the other one, and there needs to be one available at all times. So everyone working outside will need to go and sit in the crib room until either the end of the day, or an ambulance becomes available. Did you know,' he continues, 'that those two ambulances service all the coal mining and construction sites in this area, meaning that over 3,000 workers have just been stood down?'

Although I should be working, I keep ducking into the crib room to check the mood. Initially it's buzzing as nobody has any further information, but everyone is searching for some. People are then sad but excited, as they've got the afternoon off. An hour later they're bored, as they've come to realise they're stuck with nothing to do or drink. Although I don't hear anyone admit it, I get the feeling that they'd prefer to be working, so the time would pass faster. From an early age my Mum taught me to carry a book everywhere in case I get stuck waiting, and I do notice that out of the hundred-odd in the crib room, the few that are reading appear to be the most content – but I am biased, as I'd pick reading over almost anything, especially this job.

When 5 pm comes and goes, nobody's been allowed back to work and there's no sign of a bus. Around 7 pm one arrives, and those who haven't yet gotten a lift make it back to camp. Anyone who had a commute of over an hour apparently then missed out on dinner, as nobody had communicated what was happening to the kitchen, and several complaints were lodged. Since I'm now living where I work, I sat in the crib room with a book and observed until 7 pm, when it became empty, and then made it to dinner by 7.15 pm – plenty of time before it finished at 8 pm.

Day 75 – Sunday, March 24

At the morning meeting, I notice that the number on the large whiteboard that lists the days since the last lost time injury (LTI) hasn't been reset to zero, and is still climbing.

Late last year, Sammy from JRT Projects spent two weeks in hospital after a workplace injury, which was officially recorded as 'light office duties' as the mining company really likes maintaining that streak of days without an LTI, as it looks great on press releases and in shareholder reports. I have no idea, however, how anyone could spin a death into 'light office duties'.

So I ask Leon, who explains that, 'It doesn't count as a lost time injury because that bus driver who passed away won't be coming back to work. Can't lose any time when you haven't got any left.'

From others around the site I learn that the bus driver was 35, died of a massive heart attack and didn't have a pre-existing heart condition. Whenever he wasn't at work, I'm also told, he always he had a ciggy in one hand and a red wine in the other. Meaning that he'd basically smoked and drank himself to an early death.

'Three young kids too, and earlier this year they lost their house in the floods. It's all pretty fucked,' says Donk.

That bus driver was only a year older than me, and if I needed any extra inspiration to lay off the heavy drinking, well here it is. Maybe it'll inspire a few others to ease up as well.

Before leaving work, I learn his name was Paul something, and that there's a benefit happening at the Black Nugget in town tomorrow night to raise money for his family and funeral. I try to remember meeting a bus driver named Paul, and feel awful when I realise that Gary, who I met last year, is still the only bus driver I know by name, and I have no idea whether any of the fuzzy faces I can recall now belong to a dead man.

Day 76 – Monday, March 25

All day I try to get a sense of whether anyone's heading to the benefit, but it's not mentioned once. Then the moment the workday's done, nearly everyone onsite departs for the Black Nugget.

Someone has blown up and printed out a photo of a smiling Paul, and it's behind the bar, resting between the rows of spirits and the mirrored wall. I don't remember ever meeting him, and notice that while overweight, he's not outrageously obese or haggard-looking.

No specific fundraising activities appear to have been planned and while the mood is subdued, everyone is standing around in their high-vis shirts drinking, talking, smoking and gambling, much the same as any night at the pub before a rostered day off. The only differences being that we all have to work tomorrow, and we're not here for us.

There are four buckets spaced out along the bar, each with the same photo of Paul taped to them, and I notice that after paying for their drinks, most drop in the change. There's a constant stream of fifty- and twenty-dollar bills flowing in as well, and after buying a beer, I add in the change and a twenty of my own.

Ben tells me, 'Last year I was at this site where a young apprentice was taken out by a road train. It was the same then, with the buckets. We certainly look after our own.'

Instead of drinking beers at my old pace, I'm sipping and floating between conversations, more observing and listening than taking an active part. As the alcohol takes effect on the crowd, I notice them become even rowdier than the night before an RDO, but not quite at Xmas party levels.

Last night I'd looked up some stats on smoking, and was surprised to discover that less than 10 per cent of lifelong smokers get lung cancer; although male smokers are 23 times more likely to get it than non-smokers. However, I also discovered that there are plenty of other ways smoking can kill you and that half the people who smoke are eventually killed by it, and half of those before they're fifty.

Looking slowly around the large outdoor smoking area at the Black Nugget, with its tiled concrete floor, picnic tables and sparse potted palms, I count sixty smokers. It occurs to me that a quarter of them, which is exactly fifteen of the people I'm looking at right now, will probably be dead before they're fifty.

Surely the passing of Paul the bus driver has gotten others thinking along similar lines. So I insert myself into conversations, and ask as delicately as I can whether anyone now plans to ease off their smoking or drinking, and am repeatedly stared at with looks of such astonishment and revulsion, it's as if I've just grown a big, floppy penis out of my shoulder.

Then I'm told things like:

'Mate, what's the grog and cigs got to do with anything?'

'This shit bad for you, is it? Well, show me something that's this much fun that isn't.'

'When it's your turn to go, it's your turn to go. Fuck all you can do about it.'

Dale, Ben, Pando, Cliff and Jonno are all standing together, and smoking.

'I only smoke when I'm out here, at the mine,' says Jonno. 'At home, I don't touch the things.'

'I did that hypnotism thing that supposedly never fails,' says Ben. 'Then I got in the car and first thing I did was light up a smoke.'

'I'm wearing two patches right now,' Pando tells us. 'You certainly get a decent whack from each smoke when you're wearing patches.'

'Got a spare one?' asks Ben.

'Sorry, they're all back at my donga.'

'I used to only smoke bongs,' Cliff explains. 'But that's basically the only shit that shows up on the drug test.'

'My daughter caught me smoking, and I told her it's a sometimes thing Daddy does when he's stressed,' says Dale. 'She misheard me, went to school and told all the teachers I was depressed, then on parent-teacher night everyone I met kept handing me pamphlets and shit, and smiling at me like I was retarded.'

Jonno shrugs. 'I know it's bad for you, but I've got other stuff to worry about. Like finishing this fucking job on time.'

In my twenties I often drank enough to think smoking was a wonderful idea, not because I'd forget about the ill effects, but I'd just stop caring. About the next day, my future, anything. In those moments I'd be so desperate to get wasted, that I shoved whatever I could into my face. It was plenty of fun, but I remember also not liking myself much at those times and maybe that's got something to do with it. You know the consequences and don't care, because there's nothing else apart from having a good time right now to live for.

Around midnight the buckets are passed around for one last time, and I see that they're packed tight and to the top with notes. Then last drinks are called and I know I'll be fine for the breath test tomorrow morning, but I wonder if anyone will be caught out. Every night out here there's always a few groups up and drinking past midnight but they've all still got jobs, so are either well aware of their limits, or experts at avoiding the test.

In the line for the mini-bus back to camp I overhear the guy next to me say, 'What a send off. I can't remember the last time I was this pissed on a work night.'

His friend adds, 'The perfect way to remember a poor fucker who loved to get on it, but will never get that chance again.'

Dispatch No 30 – Tuesday, March 26 to Wednesday, March 27

Out with a whimper

Day 77 – Tuesday, March 26

Tomorrow I fly out for the Melbourne International Comedy Festival. Ticket sales have been sluggish, but they were also slow leading up to both the Adelaide and Perth Fringes, and they both ended up returning a modest profit. My debt is finally under $10,000, and if I break even in Melbourne, after one more swing, I'll finally be back in the black.

I'm listing the details of my upcoming Melbourne shows on different entertainment websites when Jonno appears behind me.

'What are you up to there, Xavier mate?'

'Just looking up some stuff.'

'What stuff?'

It takes me a comically long time to click all the browser windows closed.

'What's going on in Melbourne, that sort of thing,' I reply.

'So nothing to do with work then?'

'No. Just taking a short break.'

Jonno nods. There's already something very wrong with this conversation. He moves around the desk to stand in front of me. 'How are those manuals going? Nearly done?'

'Nearly.' I still haven't started them.

'May I have a look?'

'Sure.' I click some things open, making a show of looking for them while formulating an excuse.

'Dammit!' I bang the table. 'I left them back at the camp. They're on a memory stick.'

'Why would you do that?'

'Backups, ya know? For security. Also, I was hoping to finish them off while I was away.'

'Is that right?'

He drops an invoice on my desk, for some day labour we'd charged to Debitel.

'Can you tell me what's wrong here?'

I stare at it. I've prepared so many of these that I don't remember this one. 'Is it a spelling mistake?'

'I've noticed that you hardly ever make spelling mistakes, when you pull your finger out and bother doing some work.'

He's staring at me but his eyes are vacant, as if he's not seeing me and is just repeating lines from a script. 'So it's not the spelling,' he continues. 'Something much worse.'

'I'm sorry,' I tell him. 'I don't see it.'

'I suppose if you'd seen it, you would've fixed it before you sent it off for payment. Look at this number "10" right here, for the total hours worked that week.'

'Okay.'

'That number should be 100. We charge out our machine operators at $120 an hour. So the total should be $12,000. Not $1,200, which is what it says.'

'It'll only take me a second to fix, and we can send it again.' I quickly find and open up the invoice.

'Too late. They've already paid it. You just cost the company $12,000.'

Usually I'd just nod and smile, but this is really bad, and I can tell he's not done.

'You forgot to subtract the $1,200 that Debitel has paid. So it's actually only $10,800.'

'So now you know how to add up properly?' Jonno replies.

'Aren't you supposed to check all the invoices before they go out?'

'What's the point of you being here if I have to check every single little thing that you do?'

'Didn't Peter and Scott do a handshake deal with Debitel that covers all this stuff?' I ask.

'It didn't include day labour. That's always different.'

I lean back in my chair, and swivel the cursor around my screen. 'So all this extra responsibility. Does it come with a pay increase?'

'What, so you're finally paid more than Jerome?' He smiles.

I knew it. I knew that useless fuck had been getting more money than me. 'What about the hundreds of thousands I managed to get out of Debitel with all that estimating rubbish? All those invoices I cross checked? And the ones I forged? All those refunds I got from suppliers?'

'That's called doing your fucking job. Now those manuals. Can you show me anything?' His voice is getting higher with each sentence. 'I'd like to just have a quick look. You know, to check you're on the right track.'

'Sorry, no.'

We both know he's got me.

He nods again, more slowly this time, and his voice returns to normal. 'This job is just about done, and there's no more work out here for you, Xavier mate. Just park the car at the airport tomorrow will you? Leave the keys under the front seat.'

'No problem.'

This is clearly all he came over to tell me and I don't understand, if he was always going to fire me, why he needed to bother with all that other crap.

'Did you still want me to finish off those manuals?' I ask. 'And get the invoices for the month all sorted? I can do it while I'm in Melbourne.'

'I thought you'd done that already?'

'Nearly done, but not quite.' It's something else I haven't started.

'Would JRT need to pay you to do that?'

'Who works without getting paid?'

'You do all the time, and call it a comedy career.'

He laughs, and I surprise myself by laughing as well.

'Don't worry about it,' he says, and heads out to the smoking area.

I wonder how similar that conversation was to the one he had with Jerome and I wonder if this, what I'm feeling right now, the anger and confusion and embarrassment, is the same thing everyone feels after they've been fired. It's the first time it's ever happened to me.

As I pack up my desk, which takes all of three minutes, I realise that Jerome started out here before me, which means he lasted at least four months longer in this job than I did. There's some perspective for you.

I do genuinely feel awful for never starting on those manuals. That invoice was an honest mistake, and I did complete nearly every other task I was given. Those manuals though, the mountains of technical, lifeless words that nobody was ever going to read, they probably wouldn't have even taken very long. I'd thrown together similar things while working as an engineer, and it was mostly a matter of cutting, pasting then proofreading.

I want to say the reason why I didn't even attempt them is because I'm a real writer, and for me that sort of writing is like asking a professional portrait painter to do an undercoat in your bathroom. I'm not a professional writer, however, and if it was offered to me, I'd probably take that job painting the bathroom.

The real reason I didn't do it is because I knew it'd be boring, and I just couldn't be bothered. Completely unlike an adult, and entirely like a spoilt man-child.

Ah, screw it. I wanted to last six months to a year out here. I made it to five, and if the Melbourne International Comedy Festival goes okay, my debt will be manageable enough to allow me to return to being a full-time writer and comedian. Meaning I'll have gotten out of this experience exactly what I wanted. I really just wish it hadn't ended like this.

Also, with all the notes I've been taking, I've probably got enough material for a book or two and a comedy show, and isn't that what artists do? Quietly fume at all the injustices they perceive they suffer, then have the last word by writing up their version of events?

Regardless, I did just get canned, and no matter what the situation, it's never nice to be told that you're not wanted. Really, I was finished the moment I knocked back that position as second in charge.

I phone the uncle who got me this job, worried that he'll have heard about my work ethic, and it turns out he has but doesn't care.

'Don't worry about it,' he tells me. 'You were only there for the money, and I know for a fact that you did a far better job than most of the morons and plebs JRT has on the books.'

Day 78 – Wednesday, March 27

At 4 am I leave for the airport, and arrive before it's even open. Sitting alone at one of the metal picnic tables that's been welded into the courtyard of the carpark, I watch one final sunrise as the bugs feast on my bare legs.

As the sky slowly brightens, I think about all the amazing coincidences of evolution that led to this, and how far we've come as a species. About how lucky I was to be born into this particular life, to even have a shot at being an artist, and then I think about the problem of work. One of the most satisfying things a human being can do is meaningful work for a respectable wage, surrounded by a loving environment. The FIFO life is the exact opposite of that. Apart from the money, it's truly awful for every single person involved.

Despite how far we've all come, it's horrible that so many of us still end up out here, doing this extraordinarily inhuman and shit thing to ourselves, and to the planet.

When I'm away performing at comedy and fringe festivals I'm often asked if I'm morally and ethically okay with taking money from companies that clearly don't give a stuff about the environment, workers or broader society, and I have to say that the answer is a resounding, 'Fuck yes.'

It's a decision everyone should make for themselves, but as long as nobody is telling me what I can or can't say or write, then I'm always going to be okay with it.

Honestly, I like to think I'm not driven by money, but by a need to be creative and hopefully share something that makes people laugh, provokes thought and debate, and maybe even improves things. Unfortunately, good intentions don't pay for shit and after my five months out here, I now have one thing that I didn't have before I left – options.

Being an artist has been ridiculously hard all through history, and there's no way I'd be able to return to it so quickly after the Edinburgh Fringe Festival $20,000 disaster if it wasn't for the mining money. So even if that cash has come from companies who tear coal from the ground and sell it to the Chinese, I had to admit that I was thankful for it.

While I'll probably never stop feeling a little bit disgusted with myself, it's far easier to feel horrible in jeans without holes in the crotch, checked shirts that aren't stained or torn, and shoes that aren't held together by duct tape and hope. Then when it's my turn to buy a round of drinks at the bar, it's truly lovely to be able to afford something other than tap water.

At 6 am the roller doors finally rise. After checking in and going through security then into the gate lounge, I'm feeling far more sociable than usual, probably because I'm never coming back. So I join four others in high vis also about to fly out. They turn out to be actual underground coal miners and do all appear to be wearing eyeliner, which I now know is the coal that's hardest to scrub off. I ask why they got into FIFO, and every one of them says, 'For the money.'

'What other reason is there?' asks one in an orange high- vis shirt so thick with coal that it resembles the sky during a bushfire.

I explain with a smile that I was just fired, and am returning to Melbourne to be a full-time comedian.

One with a beard that's either jet black or 90 per cent coal dust asks me, 'If you don't have a real job, how are you ever going to retire?'

When I tell anyone I'm a comedian, they assume it isn't a real job and I suppose they're right, because I'm not yet making a wage I could live on out of it.

'To be honest, I don't think I'll ever retire. I mean, if retirement is finishing up your working life to follow your passion, then I retired years ago when I quit being an engineer.'

'But what about your family?' asks a miner with big, thick black eyebrows that I also suspect are packed with coal.

'I don't have one.'

'Clever,' says the one in the burnt orange shirt.

These four work two weeks on, one week off, and a pair of them have been at it for over a decade. They all have young families and are desperate to stop working FIFO, but all agree that they first want to put their children through school, pay off the house and save a bit. So instead of investing in their own dreams, like I do, these men provide for their families and then themselves. I can't decide if my choice, compared to theirs, is selfish, stupid, smarter or just different.

'Even if I didn't mind it, you're going to be sick of anything after you've done it twelve hours a day, fourteen days at a stretch, for as long as I have,' says burnt orange.

'Too right,' the others agree.

'And anyway, there's very little work around in the cities for blokes like us, who didn't finish high school and did a trade instead,' says eyebrows.

Then the conversation shifts to which bar they should head for after our flight lands as 10 am, to kick off their usual post-swing bender.

I won't be joining them, but I certainly understand why they need a drink. Every Friday afternoon in Australian cities, the bars are jammed with workers washing away another week with alcohol. A few hours of drinking to blot out around forty hours of pain, and these guys have worked around $12 \times 7 \times 2 = 168$ hours without a day off. Meaning they've got four times the pain to wash away.

Most jobs involve a swap, where you trade your time for money, which you then use to fulfil your needs and wants. For most of those I've met who work FIFO, so much of their time is taken up with it that there's little left over for anything else, and while in the industry, their lives are put on hold in exchange for a better future. Hopefully.

The most precious thing each of us has is our own meagre moments on this planet, and that's exactly what the FIFO life takes. So if you don't get out with enough cash, or with enough time to enjoy it, then you've sacrificed everything for nothing.

So what's the solution? The mining companies, despite all the crap about how much they care that's pumped out through their marketing departments and press releases, they really only exist to return the maximum possible profit to their shareholders. Like every company ever.

So FIFO work exists because the mining companies have calculated that it's the most economical way to get the work done. If it were cheaper to establish proper towns at the mining sites, as well as providing shorter shifts and swings, and other measures that improved the mental health and wellbeing of the workers, then they would. It's not, so they don't.

It's not as if the mining companies can't afford to, but they never will unless forced into it by unions and the government. Which will never happen, because the unions are being pushed out and the government wouldn't want to piss off big business, who are their real bosses.

Years from now, we're going to be cursing the legacy left by the mining industry in Australia, if we're not already.

At 8 am I'm on the plane, boarded and ready to go and I've even got a window seat. As we take off I start chatting to the FIFO seated beside me, who's heading home to meet his first grandchild, and he gives me the best advice I've ever heard.

What is it about this life that you enjoy the most?

Don't stop looking until you've found it.

Then do it as much as you can as soon as you can, because you have much less time than you think.

Otherwise, before you know it, your later will become never, and your dreams will be replaced with regrets.

Printed in Great Britain
by Amazon